Understanding Climate Change

SELECT BERNAN PRESS TITLES BY FRANK R. SPELLMAN

Environmental Science and Technology
Essentials of Environmental Engineering
The Handbook of Nature
The Handbook of Environmental Health (with Melissa Stoudt)
The Handbook of Geoscience (with Melissa Stoudt)
In Defense of Science: Why Scientific Literacy Matters (with Joan Price-Bayer)
Water Pollution Control Technology (with Nancy E. Whiting)

Understanding Climate Change

A Practical Guide

Frank R. Spellman

Lanham • Boulder • New York • London

An imprint of The Rowman & Littlefield Publishing Group, Inc.
4501 Forbes Boulevard, Suite 200, Lanham, Maryland 20706
www.rowman.com
800-865-3457; info@bernan.com

86-90 Paul Street, London EC2A 4NE

British Library Cataloguing in Publication Information Available

Library of Congress Cataloging-in-Publication Data
Names: Spellman, Frank R., author.
Title: Understanding climate change : a practical guide / Frank R. Spellman.
Description: Lanham : Bernan Press, [2021] | Includes bibliographical references and index.
Identifiers: LCCN 2021024414 (print) | LCCN 2021024415 (ebook) | ISBN 9781641434218 (paperback) | ISBN 9781641434225 (epub)
Subjects: LCSH: Climatic changes. | Climate change mitigation.
Classification: LCC QC902.8 .S64 2021 (print) | LCC QC902.8 (ebook) | DDC 551.6—dc23
LC record available at https://lccn.loc.gov/2021024414
LC ebook record available at https://lccn.loc.gov/2021024415

Contents

Introduction

> Globally, 2020 was the hottest year on record, effectively tying 2016, the previous record. Overall, Earth's average temperature has risen more than 2 degrees Fahrenheit since the 1880s. Temperatures are increasing due to human activities, specifically emissions of greenhouse gases, like carbon dioxide and methane.
>
> —NASA Goddard Institute for Space Studies, news release, January 14, 2021

Is Earth's climate changing? Are warmer times on the way? Are colder times on the way? Is the greenhouse effect affecting our climate, and if so, do we need to worry about it? Will sea levels rise—and inundate New York City? Are hurricanes more numerous and destructive? Are wildfires going to destroy the West Coast? Does the widening ozone hole portend disaster right around the corner? Do we need to eliminate cows and other ruminants because of their methane production? Is it not true that over the millennia climate change has occurred several times in Earth's past?

Are humans responsible for climatic change?

These and so many other questions related to climate change bombard us all. Newspaper headlines, magazine articles, and television/online news reports inundate us. Politicians and policy makers present us with conflicting warnings and possible solutions.

Scientists are also part of the chorus, reporting on the catastrophic harm atmospheric warming brings to the world. Some scientists state that global warming could bring record droughts, record heat waves, record smog levels, and an increasing number of forest fires. Others caution that the increasing atmospheric heat could melt the world's icecaps and glaciers, causing ocean levels to rise to the point where some low-lying island countries would disappear, while the coastlines of other nations would be drastically altered for ages—or perhaps for all time.

What's going on? We hear plenty of theories put forward by doomsayers, but are they correct?

If they are correct, what does it all mean?

Should we panic?

No. While we should be concerned, no real cause for panic exists.

Should we take some type of decisive action—should we come up with quick answers and put together a plan to fix these problems?

What really needs to be done?

We can study the facts, the issues, the possible consequences—but the key to successfully combating these issues is to stop and seriously evaluate the problems. Only through a cool-headed, scientifically intellectual, informed mindset will we be able to solve our environmental dilemma. To save our environment (and ourselves) we must develop a vision of an environmentally healthy world—an accomplishable vision. And it *is* something we can accomplish.

Another question that has merit here is, "Will we take the correct actions before it is too late?" The key words here are: "correct actions." Eventually, we may have to take some. But we do not yet know what those actions could be or should be.

Through time and experience we have learned (yes, professors and "experts" learn, too) that whether we call it global warming, global climate change (human-induced global warming, under a broader label), or an inconvenient truth, the topic is a conundrum.

Consider this: Any damage we do to our atmosphere affects the other three environmental mediums: water, soil, and biota (us—all living things). Thus, the endangered atmosphere (if it is endangered) is a major concern (a life and death concern) to all of us.

Understanding Climate Change: A Practical Guide provides explanations and answers, based on the author's research and opinion, to all the above questions and many more. This book is centered on the United States Environmental Protection Agency's (USEPA's) Climate Change Indicators in the United States. Each of these indicators are described and discussed in detail. It is my hope to address the following points:

- Distinguishing between climate change and global warming.
- Looking at Earth's climate over the millennia and in particular ice ages.
- Explaining the greenhouse effect, greenhouse gases, and their role in climate change.
- Discussing the effects of these forces now and in the near future.
- Looking at how pollution is involved.

This timely book builds on the success of several of the author's environmental science editions. Primarily written as an information source, this text is not limited in its potential for other uses. For example, while this work can be utilized by the environmental practitioner to provide valuable insight into the substance he or she works hard to collect, test, treat, and retest for its intended purpose, it can just as easily provide important information for the policy maker who may be tasked with making decisions concerning the impacts of climate change. Moreover, this book will serve a varied audience: students, lay personnel, regulators, technical experts, attorneys, business leaders, and concerned citizens. Finally, this book is clearly written and user-friendly, presented in the author's conversational style. Because when you want to communicate, it is important to speak to the reader.

Frank R. Spellman
Norfolk, Virginia

Chapter 1

Climate and the Geological Past

Before we can look at present and near-present climatic conditions and global warming, we need to look at the—cooler—past. We need to define the era we refer to when we say "the past." Tables 1.1 and 1.2 are provided to assist us in making this definition. Table 1.1 gives the entire expanse of time from Earth's beginning to present. Table 1.2 provides the sequence of geological epochs over the past sixty-five million years, as dated by modern methods. The Paleocene through Pliocene together make up the Tertiary Period; the Pleistocene and the Holocene compose the Quaternary Period.

Table 1.1. Geologic Eras and Periods

Era	*Period*	*Millions of Years BP*
Cenozoic	Quaternary	2.5–present
	Tertiary	65–2.5
Mesozoic	Cretaceous	135–65
	Jurassic	190–135
	Triassic	225–190
Paleozoic	Permian	280–225
	Pennsylvanian	320–280
	Mississippian	345–320
	Devonian	400–345
	Silurian	440–400
	Ordovician	500–440
	Cambrian	570–500
Precambrian		4,600–570

Note: BP = before the present

Table 1.2. Geological Epochs

Epoch	*Million Years Ago*
Holocene	.01–0
Pleistocene	1.6–.01
Pliocene	5–1.6
Miocene	24–5
Oligocene	35–24
Eocene	58–35
Paleocene	65–58

When we think about climatic conditions in the prehistoric past, two things generally come to mind—Ice Ages and dinosaurs. Of course, in the immense span of time that prehistory covers, those two items represent only brief moments. So let's look at what we know or what we think we know about the past, and about Earth's climate and conditions. One thing to consider—geological history shows us that the normal climate of Earth was so warm that subtropical weather reached to 60° N and S, and polar ice was entirely absent.

Only during less than about one percent of Earth's history did glaciers advance and reach as far south as what is now the temperate zone of the northern hemisphere. The latest such advance, which began about 1,000,000 years ago, was marked by geological upheaval and the advent of human life. During this time, vast ice sheets advanced and retreated, grinding their way over the continents.

A TIME OF ICE

Nearly two billion years ago, the oldest known glacial epoch occurred. A series of deposits of glacial origin in southern Canada, extending east to west about 1,000 miles, shows us that within the last billion years or so, apparently at least six major phases of massive, significant climatic cooling and consequent glaciation occurred at intervals of about 150 million years. Each lasted perhaps as long as 50 million years.

Examination of land and oceanic sediment core samples clearly indicate that in more recent times (the Pleistocene epoch to the present), many alternating episodes of warmer and colder conditions occurred over the last two million years (during the middle and early Pleistocene epochs). In the last million years, at least eight such cycles have occurred, with the warm part of the cycle lasting a relatively short interval.

During the Great Ice Age (the Pleistocene epoch), ice advances began—a series of them that at times covered over one quarter of the earth's land surface. Great sheets of ice thousands of feet thick, these glaciers moved across North America over and over, reaching as far south as the Great Lakes. An ice sheet thousands of feet thick spread over Northern Europe, sculpting the land and leaving behind lakes, swamps, and terminal moraines as far south as Switzerland. Each succeeding glacial advance was apparently more severe than the previous one. Evidence indicates that the most severe one began about 50,000 years ago and ended about 10,000 years ago. Several interglacial stages separated the glacial advances, melting the ice. Average temperatures were higher than those we experience today.

Wait a minute! Temperatures were higher than today? Yes, they were. Think about that as we proceed.

Because one-tenth of the globe's surface is still covered by glacial ice, scientists consider the earth still to be in a glacial stage. The ice sheet has been retreating since the climax of the last glacial advance, and world climates, although fluctuating, are slowly warming.

From our observations and from well-kept records, we know that the ice sheet is in a retreating stage. The records clearly show that a marked worldwide retreat of ice has occurred over the last hundred years. World famous for its 50 glaciers and 200 lakes, Glacier National Park in Montana does not present the same visual experiences it did a hundred years ago. In 1937, a ten-foot pole was put into place at the terminal edge of one of the main glaciers. The sign is still in place, but the terminal end of the glacier has retreated several hundred feet back up the slope of the mountain. Swiss resorts built during the early 1900s to offer scenic glacial views now have no ice in sight. Theoretically, if glacial retreat continues, melting all of the world's ice supply, sea levels would rise more than 200 feet, flooding many of the world's major cities. New York and Boston would become aquariums.

The question of what causes ice ages is one scientists still grapple with. Theories range from changing ocean currents to sunspot cycles. There's one fact of which we are absolutely certain, however: an ice age event occurs because of a change in Earth's climate. But what could cause such a drastic change?

Climate results from uneven heat distribution over Earth's surface. It is caused by the earth's tilt—the angle between the earth's orbital plane around the sun and its rotational axis. This angle is currently 23.5 degrees, but it has not always been that. The angle, of course, affects the amount of solar energy that reaches the earth, and where it falls. The heat balance of the earth, which is driven mostly by the concentration of carbon dioxide (CO_2) in the atmosphere, also affects long-term climate. If the pattern of solar radiation changes or if the amount of carbon dioxide changes, climate change can result.

Evidence (primarily from soil core samples and topographical formations) tells us that change in climate includes events such as periodic ice ages characterized by glacial and interglacial periods. Long glacial periods lasted up to 100,000 years; temperatures decreased about 9°F, and ice covered most of the planet. Short periods lasted up to 12,000 years, with temperatures decreasing by 5°F and ice covering 40° north latitude and above. Smaller periods (e.g., the "Little Ice Age," which occurred from about 1000 to 1850 CE) had about a 3°F drop in temperature. *Note*: Despite its name, the Little Ice Age was a time of severe winters and violent storms, not a true glacial period. These ages may or may not be significant but consider that we are presently in an interglacial period and that we may be reaching its apogee.

Let's look at the effects of ice ages (i.e., effects we think we know about). Changes in sea levels could occur. Sea level could drop by about 100 meters in a full-blown ice age, exposing the continental shelves. Increased deposition during melt would change the composition of the exposed continental shelves. Less evaporation would change the hydrological cycle. Significant landscape changes could occur—on the scale of the Great Lakes formation. Drainage patterns throughout most of the world and topsoil characteristics would change. Flooding on a massive scale could occur. How these changes would affect you depends on whether you live in Northern Europe, Canada, the Pacific Northwest, around the Great Lakes, or near a seashore.

To generate a full-blown ice age (massive ice sheet covering most of the globe), scientists point out that certain periodic or cyclic events or happenings must occur. Periodic fluctuations would have to affect the solar cycle, for instance; however, we have no definitive proof that this has ever occurred.

Another theory speculates that periods of volcanic activity could generate masses of volcanic dust that would block or filter heat from the sun, thus cooling down the earth. Some speculate that the carbon dioxide cycle would have to be periodic or cyclic to bring about periods of climate change. There is reference to a so-called Factor 2 reduction, causing a 7°F temperature drop worldwide. Others speculate that another global ice age could be brought about by increased precipitation at the poles due to changing orientation of continental land masses. Others theorize that a global ice age would result if the mean temperatures of ocean currents decreased.

Speculation aside, what are the most probable causes of ice ages on Earth? According to the Milankovitch hypothesis, ice age occurrences are governed by a combination of factors: (1) the earth's change of altitude in relation to the Sun (the way it tilts in a 41,000-year cycle and at the same time wobbles on its axis in a 22,000-year cycle), making the time of its closest approach to the sun come at different seasons; and (2) the 92,000-year cycle of eccentricity in its orbit round the sun, changing it from an elliptical to a near circular orbit, the most severe period of an ice age coinciding with the approach to circularity.

So, what does all this mean? We know that ice ages occurred—we know that they caused certain things to occur (e.g., formation of the Great Lakes), and although there is a lot we do not know, we recognize the possibility of recurrent ice ages. Lots of possibilities exist. Right now, no single theory is sound, and doubtless many factors are involved. Keep in mind that the possibility does exist that we are still in the Pleistocene Ice Age. It may reach another maximum in 60,000 plus years or so.

REFERENCES AND RECOMMENDED READING

Crowley, T. J., and G. R. North. 1996. Abrupt Climate Change and Extinction Events in Earth's History. *Science* 240.

Chapter 2

The Atmosphere[1]

Before we dive into climate science and climate change, there is one more foundational area to cover—the atmosphere.

Shakespeare likened it to a majestic overhanging roof (constituting the transition between its surface and the vacuum of space); others have likened it to the skin of an apple. Both these descriptions of our atmosphere are fitting, as is its being described as the earth's envelope, veil, or gaseous shroud. The atmosphere is more like the apple skin, however. This thin skin, or layer, contains the life-sustaining oxygen (21 percent) required by all humans and many other life forms; the carbon dioxide (0.03 percent) so essential for plant growth; the nitrogen (78 percent) needed for chemical conversion to plant nutrients; the trace gases such as methane, argon, helium, krypton, neon,

DID YOU KNOW?

While the gases discussed here are important to maintaining life as we know it on Earth, it is water vapor (in conjunction with airborne particles, obviously) that is essential for the stability of Earth's ecosystem. This water vapor–particle combination interacts with the global circulation of the atmosphere and produces the world's weather, including clouds and precipitation.

1. Adapted from F. R. Spellman, *The Science of Air*, 2nd ed., Boca Raton, FL: CRC Press, 2007.

xenon, ozone, and hydrogen; and varying amounts of water vapor and airborne particulate matter. Life on Earth is supported by this atmosphere, solar energy, and other planets' magnetic fields.

What it takes for Earth (or any other planet) to hold an atmosphere is a high mass and low temperature. Well, some would argue that what is really needed to hold Earth's atmosphere in place is gravity. And this is true, of course, but gravity of Earth is attributed to its mass. Because of Earth's large mass relative to that of the moon, the escape velocity of any particle on Earth is 7.1 miles a second, whereas on the moon, the escape velocity is 1.5 miles a second. In simple terms what this means is that if a particle pointed its nose out to the universe and traveled at 7.1 miles per second it could escape Earth's gravitational hole. The temperature of a planet is important because the higher the temperature, the faster "air" molecules travel and collide with one another, and if they are excited enough by the energy input of a high temperature, they might reach escape velocity and leak off into space.

Gravity holds about half the weight of a uniform mixture of these gases in the lower 18,000 feet of the atmosphere; approximately 98 percent of the material in the atmosphere is below 100,000 feet.

Atmospheric pressure varies from 1000 millibars (mb) at sea level to 10 mb at 100,000 feet. From 100,000 to 200,000 feet the pressure drops from 9.9 mb to 0.1 mb and so on.

The atmosphere is considered to have a thickness of 40 to 50 miles; however, here we are primarily concerned with the **tropo-**

DID YOU KNOW?

The troposphere, stratosphere, mesosphere, and thermosphere act together as a giant safety blanket. They keep the temperature on the earth's surface from dipping to extreme icy cold that would freeze everything solid, or from soaring to blazing heat that would burn up all life.

sphere, the part of the earth's atmosphere that extends from the surface to a height of about 27,000 feet above the poles, about 36,000 feet in mid latitudes, and about 53,000 feet over the equator. Above the troposphere is the **stratosphere**, a region that increases in temperature with altitude (the warming is caused by absorption of the sun's radiation by ozone) until it reaches its upper limit of 260,000 feet.

The troposphere is where people, plants, animals, and insects live and depend on its thin layer of gases. Weather is generated in this region. Within the troposphere is the phenomena of jet streams. The jet streams are very influential in determining weather patterns: they undulate snake-like from north to south across North America. If they go northward, the weather becomes warmer in the south, if it moves southward, the weather becomes warmer in the north.

The stratosphere begins around the eight-mile point and reaches 31 miles into the sky. In the rarified air of the stratosphere, the significant gas is ozone (life-protecting ozone O_3—not to be confused with pollutant ozone), which is produced by the intense ultraviolet radiation from the sun. In quantity, the total amount of ozone in the atmosphere is so small that if it were compressed to a liquid layer over the globe at sea level, it would have a thickness of less than 3/16 inch.

In the stratosphere, middle sized wavelengths of ultraviolet (UV) radiation are absorbed and used when O_1 and O_2 combine to make O_3—Ozone. Ozone's ability to absorb (UV) radiation increases the energy level contained in the stratosphere and increases heat level.

Ozone contained in the stratosphere can also impact (add to) ozone in the troposphere. Normally, the troposphere contains about 20 parts per billion of ozone. On occasion, however, via the jet stream, this concentration can increase to 5–10 times higher than average.

In our discussion of Earth's atmosphere in this book the focus is on the troposphere and stratosphere because these two layers

DID YOU KNOW?

Oceans and the atmosphere are constantly interacting—exchanging heat, water, gases, and particles. As the atmosphere warms, the ocean absorbs some of this heat. The amount of heat stored by the ocean affects the temperature of the ocean both at the surface and at great depths. Warming of the earth's oceans can affect and change the habitat and food supplies for many kinds of marine life—from plankton to polar bears. The oceans also absorb carbon dioxide from the atmosphere. Once it dissolves in the ocean, carbon dioxide reacts with sea water to form carbonic acid. As people put more carbon dioxide into the atmosphere, the oceans absorb some of this extra carbon dioxide, which leads to more carbonic acid. An increasingly acidic ocean can have negative effects on animal life, such as coral reefs (USEPA 2010).

directly impact life as we know it and are or can be heavily influenced by pollution and its effects.

A JEKYLL-AND-HYDE VIEW OF THE ATMOSPHERE

When non-city dwellers look up into that great natural canopy above our heads, they see many features provided by our world's atmosphere that we know and enjoy: the blueness and clarity of the sky, the color of a rainbow, the spattering of stars reaching every corner of blackness, the magical colors of a sunset. The air they breathe carries the smell of ocean air, the refreshing breath of clean air after a thunderstorm, and the beauty contained in a snowflake.

But the atmosphere sometimes presents another face—Mr. Hyde's face. The terrible destructiveness of a hurricane, a tornado, monsoon, typhoon, or a hailstorm, the wearying monotony of winds carrying dusts and rampaging windstorms carrying fire up a hillside—these are some of the terrifying aspects of the other face of the disturbed atmosphere.

The atmosphere can also present a Hyde-like face whenever humans are allowed to pour their filth (pollution) into it. The patches of the atmosphere not blocked by buildings has made the visible sky dirty yellow-brown or, at best, a sickly pale blue and masked the stars due to pollution rising from enterprises.

Fortunately, Earth's atmosphere is self-healing. Air-cleaning is provided by clouds and the global circulation system, which constantly purge the air of pollutants. When air pollutants overload nature's way of rejuvenating its systems to their natural state we are faced with the repercussions that can be serious, even life-threatening.

ATMOSPHERIC PARTICULATE MATTER

Along with gases and water vapor Earth's atmosphere is literally a boundless arena for particulate matter of many sizes and types. Atmospheric particulates vary in size from 0.0001 to 10,000 microns. Particulate size and shape have a direct bearing on visibility. For example, a spherical particle in the 0.6-micron range can scatter light in all directions, reducing visibility.

The types of airborne particulates in the atmosphere vary widely, with the largest sizes derived from volcanoes, tornados, waterspouts, embers from forest fires, and seed parachutes, spider webs, pollen, soil particles, and living microbes.

The smaller particles (the ones that scatter light) include fragments of rock, salt and spray, smoke, and particles from forested areas. The largest portions of airborne particulates are invisible. They are formed by the condensation of vapors, chemical reactions, photochemical effects produced by ultraviolet radiation, and ionizing forces that come from radioactivity, cosmic rays, and thunderstorms.

Airborne particulate matter is produced either by mechanical weathering, breakage, and solution or by the vapor-to-condensation-to-crystallization process (typical of particulates from a furnace of a coal-burning power plant).

As you might guess, anything that goes up must eventually come down. This is typical of airborne particulates also. Fallout of particulate matter depends, obviously, mostly on their size—less obviously, on their shape, density, weight, airflow, and injection altitude. The residence time of particulate matter also is dependent on the atmosphere's cleanup mechanisms (formation of clouds and precipitation) that work to remove them from their suspended state.

Some large particulates may only be airborne for a matter of seconds or minutes with intermediate sizes able to stay afloat for hours or days. The finer particulates may stay airborne for a much longer duration: for days, weeks, months, and even years.

Particles play an important role in atmospheric phenomena. For example, particulates provide the nuclei upon which ice particles are formed, cloud condensation forms, and for condensation to take place. Obviously, the most important role airborne particulates play is in cloud formation. Simply put, without clouds life as we know it would be much more difficult and cloud bursts that eventually erupted would cause such devastation it is hard to imagine or contemplate.

The situation just described could also result whenever massive forest fires and volcanic action takes place. These events would release a superabundance of cloud condensation nuclei, which would overseed the clouds, causing massive precipitation to occur. If natural phenomena such as forest fires and volcanic eruptions can overseed clouds and cause massive precipitation, then what effect would result from man-made pollutants entering the atmosphere at unprecedented levels?

Read on for answers.

REFERENCES AND RECOMMENDED READING

Anthes, R. A., J. J. Cahir, A. B. Frasier, and H. A. Panofsky. 1984. *The Atmosphere*, 3rd ed. Columbus, OH: Charles E. Merrill Publishing.

Ingersoll, A. P. 1983. The Atmosphere. *Scientific American* 249(3): 162–174.

Spellman, F. R. 2007. *The Science of Air*, 2nd ed. Boca Raton, FL: CRC Press.

USEPA. 2016. Climate Indicators. EPA 450-f-16-071. Accessed June 19, 2019, https://www.epa.gov/climate-indicators.

Chapter 3

Climate Change Science, Global Warming, and Climate Change Indicators

> Humanity is conducting an unintended, uncontrolled, globally pervasive experiment whose ultimate consequences could be second only to nuclear war. The earth's atmosphere is being changed at an unprecedented rate by pollutants resulting from human activities, inefficient and wasteful fossil fuel use and the effects of rapid population growth in many regions. These changes are already having harmful consequences over many parts of the globe.
>
> —Toronto Conference Statement, June 1988

It is essential to understand climate science in order to follow the rest of this book. So, this chapter focuses on the basics of climate science as it applies to what is affecting today and tomorrow's climate. These basics set the foundation for the material that follows—climate change indicators.

THE HEAVY HAND OF HUMANS

Humans have been altering their environment in dramatic fashion, especially since the advent of the Industrial Revolution (circa 1760). In this section, we focus on human activities that profoundly affect the environment. These activities are not secret or mysterious; in fact, they are obvious—and most of us take part in some of these activities daily. As Graedel and Crutzen (1989) put it, the activities of humans are changing our atmosphere. To summarize these activities: Humans' (1) industrial activities emit a variety of atmospheric pollutants; (2) practice of burning large quantities of fossil fuel introduces pollutants into the atmosphere; (3) transportation practices emit pollutants into the atmosphere; (4) mismanagement and alteration of land surfaces (deforestation) lead to atmospheric problems; (5) practice of clearing and burning massive tracts of vegetation produces atmospheric contaminants; and (6) agricultural practices that produce chemicals such as methane impact the atmosphere. These human-made alterations to Earth's atmosphere have produced profound effects, including increased acid precipitation, localized smog events, greenhouse gases, ozone depletion, and increased corrosion of materials induced by atmospheric pollutants.

We should understand the human-made mechanisms at work destroying our environment, and what we are collectively doing to our environment—and we must be aware that our environment is finite, not inexhaustible or indestructible. Our environment *can* be destroyed. We must also clearly identify and understand both the causal and the remedial factors involved. Recognizing one particular salient point is absolutely essential: Life on Earth and the nature of Earth's atmosphere are connected—literally chain-linked together. The atmosphere drives Earth's climate and ultimately determines its suitability for life. We must work to preserve the quality of our atmosphere.

In the following sections, we discuss those issues relevant to environmental pollution of our atmosphere and air quality on Earth. We discuss global warming, acid precipitation, photochemical smog, and stratospheric ozone depletion.

GLOBAL WARMING

What is global warming? Is it a long-term rise in the average temperature of Earth? This appears to be the case, even though the geological record shows abrupt climate changes occur from time to time (Crowley and North, 1996). Here's another question, one many people use to question the validity of the concept of global warming as an environmental hazard: Is global warming actually occurring? The answer to this accompanying question is of enormous importance to all life on Earth—and is the subject of intense debate throughout the globe. Again, all the debate for the occurrence of global warming can't dispute the historical record that points out that measurements made in central England, Geneva, and Paris from about 1700 until the present indicate a general downward trend in surface temperature (Thompson, 1995). However, recent records point to the opposite—global temperatures have been on the rise.

Now, before moving on we need to make a clear distinction between the terms *global warming* and *global climate change*. Yes, there is a difference between the two. **Global warming** refers to the observed warming of the planet due to human-caused emissions of greenhouse gases. **Climate change** refers to all of the various long-term changes in our climate, including extreme weather, sea-level rise, and acidification of our oceans.

For the sake of discussion, let's assume that global warming is occurring. With this assumption in place, we must ask other questions, ones that deal with why, how, and what. (1) Why is global warming occurring? (2) How can we be sure it is occurring? (3) What will be the ultimate effects? (4) What can and are we going to do about it? These questions are difficult to answer. The real danger is that we may not be able to definitively answer these questions before it is too late—when we've reached the point when the process has progressed beyond the power of humans to effect prevention or mitigation. Are we not to take precautionary actions now instead of later—much later, when it is too late? If we wait to take action, we may find that mitigation becomes harder, more expensive, and impossible to enact.

Exactly what is the nature of the problem of global warming and climate change? Hang on. We may not provide all the answers, but we are about to launch into a discussion of the entire phenomenon and its potential impact on Earth.

GLOBAL WARMING VS. CLIMATE CHANGE

What is the difference between "climate change" and "global warming"—is there really a difference? Yes, there is a distinct difference, just as there is a distinct difference between "weather" and "climate." This is the case even though "climate change" and "global warming" are often used interchangeably.

Global warming refers to the long-term warming of the planet since the Industrial Revolution, and most notably since the late 1970s, due to the increase in fossil fuel emissions. According to NASA (2019): worldwide since 1880, the average surface temperature has gone up by about 1°C (about 2°F), relative to the mid-twentieth-century baseline (of 1951–80). This is on top of about an additional 0.15°C of warming from between 1750 and 1880.

Climate change is different from global warming in that it encompasses the broad range of global phenomena created predominantly by burning fossil fuels, which add heat-trapping gases to Earth's atmosphere. These phenomena include the increased temperature trends described by global warming, but also encompass changes such as sea-level rise; ice mass loss in Greenland, Antarctica, the Arctic, and mounted glaciers worldwide; shifts in flower/plant blooming; and extreme weather events (NASA, 2019).

DID YOU KNOW?

Climate change can have broad effects on biodiversity (the number and variety of plant and animal species in a particular location). Although species have adapted to environmental change for millions of years, a quickly changing climate could require adaption on larger and faster scales than in the past. Those species that cannot adapt are at risk of extinction. Even the loss of a single species can have cascading effects because organisms are connected through food webs and other inactions (USEPA, 2010).

THE GREENHOUSE EFFECT

The **greenhouse effect** makes life on Earth as we know it possible. The basic science of the greenhouse effect has been understood for more than a century. To understand the greenhouse effect, here's an explanation most people (especially gardeners) are familiar with. In a garden greenhouse, the glass walls and ceilings are largely transparent to shortwave radiation from the sun, which is absorbed by the surfaces and objects inside the greenhouse. Once absorbed, the radiation is transformed into longwave (infrared) radiation (heat), which is radiated back from the interior of the greenhouse. But the glass does not allow the longwave radiation to escape, instead absorbing the warm rays. With the heat trapped inside, the interior of the greenhouse becomes much warmer than the air outside.

With Earth and its atmosphere, much the same effect takes place (see figure 3.1). The shortwave and visible radiation that reaches Earth is absorbed by the surface as heat. The long heat waves are then radiated back out toward space, but the atmosphere instead absorbs many of them. This is a natural and balanced process, and indeed is essential to life on Earth. The problem comes when changes in the atmosphere radically alter the amount of absorption, and therefore the amount of heat retained. Scientists, in recent decades, speculate that this may have been happening as various air pollutants have caused the atmosphere to absorb more heat.

That this phenomenon takes place at the local level with air pollution, causing heat islands in and around urban centers, is not questioned. The main contributors to this effect are the greenhouse gases: water vapor, carbon dioxide, carbon monoxide, methane, volatile organic compounds (VOCs), nitrogen oxides, chlorofluorocarbons (CFCs), and surface ozone. These gases delay the escape of infrared radiation from Earth into space, causing a general climatic warming. Note that scientists stress that this a natural process—indeed, Earth would be 33°C cooler than it is presently if the "normal" greenhouse effect did not exist (Hansen et al., 1986).

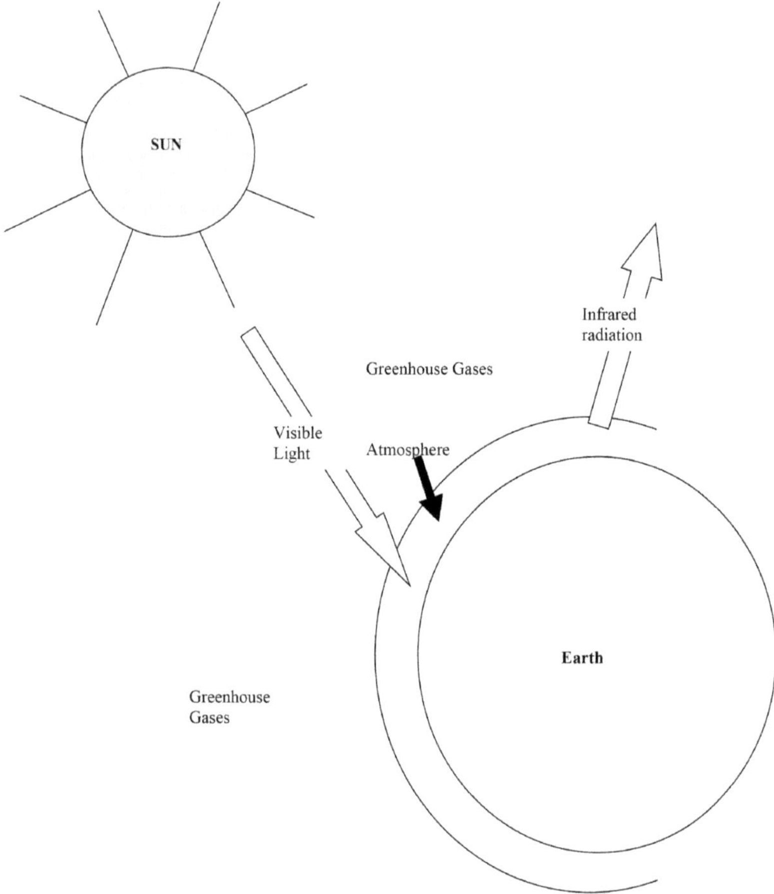

Figure 3.1. The Greenhouse Effect

The problem with Earth's greenhouse effect is that human activities are now rapidly intensifying this natural phenomenon, which may lead to global warming. There is much debate, confusion, and speculation about this potential consequence. Scientists are not entirely sure of, or agree about whether the recently perceived worldwide warming trend is because of greenhouse gases, from some other cause, or whether it is simply a wider variation in the normal heating and cooling trends they have been studying. But if it continues unchecked, the process may lead to significant global warming,

DID YOU KNOW?

Although some forests may derive near-term benefits from an extended growing season, climate change is expected to encourage wildfires by extending the length of the summer fire season. Longer periods of hot weather could stress trees and make them more susceptible to wildfires, insect damage, and disease. Climate change has likely already increased the size and number of forest fires, insect outbreaks, and tree deaths, particularly in Alaska and the West. The area burned in western U.S. forests from 1987 to 2003 is almost seven times larger than the area burned from 1970 to 1986. In the last 30 years, the length of the wildfire season in the West has increased by 78 days (USEPA, 2010).

with profound effects. The human impact on the greenhouse effect is real: it has been measured and detected. The rate at which the greenhouse effect is intensifying is now more than five times what it was during the last century (Hansen et al., 1989).

THE GREENHOUSE EFFECT AND GLOBAL WARMING

Those who support the theory of global warming base their assumptions on humans' altering the normal greenhouse effect, which provides necessary warmth for life, keeping the planet as much as 60°F warmer than it otherwise would be—ideal for humans. They blame human activities (burning of fossil fuels, deforestation, and use of certain aerosols and refrigerants) for the increased quantities of greenhouses gases. These gases have increased the amounts of heat trapped in Earth's atmosphere, gradually increasing the temperature of the whole globe.

Many scientists note that (based on recent or short-term observation) the last decade has been the warmest since temperature recordings began in the late nineteenth century and that the more general rise in temperature in the last century has coincided with the Industrial Revolution, with its accompanying increase in the use of fossil fuels. Other evidence supports the global warming theory. For example, in the Arctic and Antarctica we see evidence of receding ice and snow cover.

Taking a long-term view, scientists look at temperature variations over thousands or even millions of years. Having done this they cannot definitively show that global warming is anything more than a short-term variation in Earth's climate. They base this assumption on historical records that show Earth's temperature does vary widely, growing colder with ice ages and then warming again. The 1980s saw nine of the twelve warmest temperatures ever recorded up to that time, and the Earth's average surface temperature has risen approximately 0.6°C (1°F) in the last century (USEPA, 1995). But at the same time, that same decade also saw three of the coldest years: 1984, 1985, and 1986.

But let's assume that we are indeed seeing long-term global warming. If this is the case, we must determine what is causing it. But here, we face a problem. Scientists cannot be sure of greenhouse effect's causes. The global warming may simply be part of a much longer trend of warming since the last ice age. Though much has been learned in the past two centuries of science, little is actually known about the causes of the worldwide global cooling and warming that have sent Earth through a succession of major ice ages and smaller ones. We simply don't have the enormously long-term data to support our theories.

Factors Involved with Global Warming/Cooling

Right now, scientists are able to point to six factors that could be involved in long-term global warming and cooling.

1. Long-term global warming and cooling could result if changes in Earth's position relative to the sun occur (Earth's orbit around the sun), with higher temperatures when the two are closer together and lower when farther apart.

2. Long-term global warming and cooling could result if major catastrophes (meteor impacts or massive volcanic eruptions) occur that throw pollutants into the atmosphere that can block out solar radiation.
3. Long-term global warming and cooling could result if changes in **albedo** (reflectivity of Earth's surface) occur. If the earth's surface were more reflective, for example, the amount of solar radiation radiated back toward space instead of absorbed would increase, lowering temperatures on Earth.
4. Long-term global warming and cooling could result if the amount of radiation emitted by the sun changes.
5. Long-term global warming and cooling could result if the shape and relationship of the land and oceans change.
6. Long-term global warming and cooling could result if the composition of the atmosphere changes.

"If the composition of the atmosphere changes"—this possibility, of course, relates directly to our present concern: Have human activities had a cumulative impact large enough to affect the total temperature and climate of Earth? We are not certain, right now.

HOW IS CLIMATE CHANGE MEASURED?

Worldwide, scientists are trying to establish ways to test or measure whether or not greenhouse-induced global warming is occurring. Scientists are currently looking for signs that collectively are called a greenhouse "signature" or "footprint." If it is occurring, eventually it will be obvious to everyone—but what we really want is clear advance warning. Thus, scientists are currently attempting to collect and then decipher a mass of scientific evidence to find those signs to give us clear advance warning. According to Franck and Brownstone (1992), these signs are currently believed to include changes in:

- **global temperature patterns**, with continents being warmer than oceans; lands near the Arctic warming more than the tropics; and the lower atmosphere warming, while the higher stratosphere becomes cooler.
- **atmospheric water vapor**, with increasing amounts of water evaporating into the air as a result of the warming, more in the tropics than in the higher latitudes. Since water vapor is a "greenhouse gas," this would intensify the warming process.
- **sea surface temperature**, with a fairly uniform rise in the temperature of oceans at their surface and an increase in the temperature differences among oceans around the globe.
- **seasonality**, with changes in the relative intensity of the seasons, with the warming effects especially noticeable during the winter and in higher latitudes.

In a measured, scientific way, these signs give a general overview of some of the changes that would be expected to occur with global warming. Note, however, that from a viewpoint of life on Earth, changes resulting from long-term global warming would be drastic—profoundly serious. Probably the most dramatic—and the effect with the most far-reaching results—would be sea-level rise.

Global Warming and Sea-Level Rise

The rise in sea level is already underway, and with it will come increased storm damage, pollution, and subsidence of coastal lands.

"Rise in sea level is already underway"? Absolutely. Consider the following information taken from USEPA's 2017 report, *The Probability of Sea Level Rise*.

1. Global warming is most likely to raise sea levels 15 cm by the year 2050 and 34 cm by the year 2100. There is also a 10 percent chance that climate change will contribute 30 cm by 2050 and 65 cm by 2100. These estimates do not include sea-level rise caused by factors other than greenhouse warming.

2. There is a 1 percent chance that global warming will raise sea level 1 meter in the next 100 years and 4 meters in the next 200 years. By the year 2200, there is also a 10 percent chance of a 2-meter contribution. Such a large rise in sea level could occur either if Antarctic ocean temperature warms 5 degrees C and Antarctic ice streams respond more rapidly than most glaciologists expect, or if Greenland temperatures warm by more than 10 degrees C. Neither of these scenarios is likely.
3. By the year 2100, climate change is likely to increase the rate of sea-level rise by 4.1 mm/yr. There is also a 1-in-10 chance that the contribution will be greater than 10 mm/yr, as well as a 1-in-10 chance that it will be less than 1 mm/yr.
4. Stabilizing global emissions in the year 2050 would be likely to reduce the rate of sea-level rise by 28 percent by the year 2100, compared with what it would be otherwise. These calculations assume that we are uncertain about the future trajectory of greenhouse gas emissions.
5. Stabilizing emissions by the year 2025 could cut the rate of sea-level rise in half. If a high global rate of emissions growth occurs in the next century, sea level is likely to rise 6.2 mm/yr by 2100; freezing emissions in 2025 would prevent the rate from exceeding 3.2 mm/yr. If less emissions growth were expected, freezing emissions in 2025 would cut the eventual rate of sea-level rise by one-third.
6. Along most coasts, factors other than anthropogenic climate change will cause the sea to rise more than the rise resulting from climate change alone. These factors include compaction and subsidence of land, groundwater depletion, and natural climate variations. If these factors do not change, global sea level is likely to rise 45 cm by the year 2100, with a 1 percent chance of a 112 cm rise. Along the coast of New York, which typifies the United States, sea level is likely to rise 26 cm by 2050 and 55 cm by 2100. There is also a 1 percent chance of a 55 cm rise by 2050 and a 120 cm rise by 2100.

Along with the EPA's findings reported above, additional lines of evidence corroborate that global mean sea level has been rising during at least the last 100 years. According to Broecker (1987), this evidence is apparent in tide gauge records; erosion of 70 percent of the world's sandy coasts and 90 percent of America's sandy beaches; and the melting and retreat of mountain glaciers. Edgerton (1991) points out that the correspondence between the two curves of rising global temperatures and rising sea levels during the last century appears to be more than coincidental.

Major uncertainties are present in estimates of future sea-level rise. The problem is further complicated by our lack of understanding of the mechanisms contributing to relatively recent rises in sea level. In addition, different outlooks for climatic warming dramatically affect estimates. In all this uncertainty, one thing is certain: estimates of sea-level rise will undergo continual revision and refinement as time passes and more data is collected.

Major Effects of Sea-Level Rise: Physical and Human

With increased global temperatures, global sea-level rise will occur at a rate unprecedented in human history (Edgerton, 1991). Changes in temperature and sea level will be accompanied by changes in salinity levels. For example, a coastal freshwater aquifer is influenced by two factors: pumping and mean sea level. In pumping, if withdrawals exceed recharge, the water table is drawn down and saltwater penetrates inland. With mean sea level, the problem occurs if sea level rises and the coastline moves inland, reducing aquifer area. Additional problems brought about by changes in temperature and sea level are seen in tidal flooding, oceanic currents, biological processes of marine creatures, in runoff and landmass erosion patterns, and saltwater intrusion.

Consider the direct physical effects of sea-level rise on a coastal beach system. At current rates of sea-level rise of 1 to 2 mm/year, significant coastal erosion is already produced. Two major factors contribute to beach erosion. First, deeper coastal waters enhance wave generation, thus increasing their potential for overtopping barrier islands. Second, shorelines and beaches will attempt to establish new equilibrium positions according to what is known as the Bruun rule; these adjustments will include a recession of shoreline and a decrease in shore slope (Bruun, 1962; 1986).

Along with the physical effects of sea-level rise, in one way or another, directly or indirectly, accompanying effects have a direct human side, especially concerning settlements and the infrastructure that accompanies them: highways, airports, waterways, water supply and wastewater treatment facilities, landfills, hazardous waste storage areas, bridges, associated maintenance systems. Sea-level rise could also cause intrusion of saltwater into groundwater suppliers (Edgerton, 1991).

To point out that this infrastructure will be placed under tremendous strain by a rising sea level coupled with other climatic change is to understate the possible consequences. Indeed, the impact on infrastructure is only part of the direct human impact. For example, there is widespread agreement among scientists that any significant change in world climate resulting from warming or cooling will (1) disrupt world food production for many years; (2) lead to a sharp increase in food prices; and (3) cause considerable economic damage.

Warnings from 1997

The headlines we see in the paper sound authoritative: "2020 Was the Warmest Year on Record," "Scientists Discover Ozone Hole Has Grown," "Record Quantities of Carbon Dioxide Detected in Atmosphere." Other reports indicate we are undergoing a warming trend, but conflicting reports abound. This section discusses what we think we know about climate change.

Two environmentally significant events took place late in 1997: El Niño's return and the Kyoto Conference on Global Warming and Climate Change. News reports blamed El Niño for just about anything that had to do with weather conditions throughout the world. Some incidents were indeed related to El Niño or generated by it: the out-of-control fires, droughts, floods, the stretches of dead coral with no sign of fish in the water, and few birds around certain Pacific atolls. The devastating storms that struck the west coasts of South America, Mexico, and California were also probably connected to El Niño. El Niño's effect on the 1997 hurricane season, one of the mildest on record, is not in question, either.

Does a connection exist between El Niño and global warming or global climate change? On December 7, 1997, the Associated Press reported that while delegates at the global climate conference in Kyoto haggled over greenhouse gases and emission limits, a compelling question emerged: "Is global warming fueling El Niño?" Nobody knows for sure because we need more information than we have today. The data we do have, however, suggests that El Niño is getting stronger and more frequent.

Some scientists fear that the increasing frequency and intensity of El Niños (records show that two of the last century's three worst El Niños came in 1982 and 1997) may be linked to global warming. At the Kyoto Conference, experts said the hotter atmosphere is heating up the world's oceans, setting the stage for more frequent and extreme El Niños. Weather-related phenomena seem to be intensifying throughout the globe. Can we be sure that this is related to global warming yet? No. Without more data, more time, more science (real science), we cannot be sure.

According to the Associated Press coverage of the Kyoto Conference, scientist Richard Fairbanks reported that he found startling evidence of our need for concern. During two months of scientific experiments conducted in autumn 1997 on Christmas Island, the world's largest atoll in the Pacific Ocean, he witnessed a frightening scene. The water surrounding the atoll was 7°F higher than average for the time of year, which upset the balance of the environmental system. According to Fairbanks, 40 percent of the coral was dead, the warmer water had killed off or driven away fish, and the atoll's once-plentiful bird population was almost completely gone.

No doubt, El Niños are having an acute impact on the globe; however, we do not know if these events are caused by or intensified by global warming. *USA Today* (December, 1997) discussed the results of a report issued by the Intergovernmental Panel on Climate

Change. They interviewed Jerry Mahlman of the National Oceanic and Atmospheric Administration (NOAA) and Princeton University, and presented the following information about what most scientists agree on:

- There is a natural greenhouse effect and scientists know how it works; without it, Earth would freeze.
- Earth undergoes normal cycles or warming and cooling on grand scales. Ice ages occur every 20,000 to 100,000 years.
- Globally, average temperatures have increased 1°F in the past 100 years, within the range that might occur normally.
- The level of man-made carbon dioxide in the atmosphere has risen 30 percent since the beginning of the Industrial Revolution and is still rising.
- Levels of man-made carbon dioxide will double in the atmosphere over the next 100 years, generating a rise in global average temperatures of about 3.5°F (larger than the natural swings in temperature that have occurred over the past 10,000 years).
- By 2050, temperatures will rise much higher in northern latitudes than the increase in global average temperatures. Substantial amounts of northern sea ice will melt, and snow and rain in the northern hemisphere will increase.
- As the climate warms, the rate of evaporation will rise, further increasing warming. Water vapor also reflects heat back to Earth.

The main contributors to this effect are the greenhouse gases: water vapor, carbon dioxide, carbon monoxide, methane, volatile organic compounds (VOCs), nitrogen oxides, chlorofluorocarbons (CFCs), and surface ozone. These gases cause a general climatic warming by delaying the escape of infrared radiation from Earth into space.

Supporters of the global warming theory assume that human activities are significantly altering Earth's normal and necessary greenhouse effect. The human activities they blame for this increase of greenhouse gases include burning of fossil fuels, deforestation, and use of certain aerosols and refrigerants. Trying to pin down definitively whether or not changing our anthropogenic activities could have any significant effect on lessening global warming, though, is difficult. Scientists look at temperature variations over thousands and even millions of years, taking a long-term view of Earth's climate (see chapter 2). The variations in Earth's climate are wide enough that they cannot definitively show that global warming is anything more than another short-term variation. Historical records have shown that the earth's temperature does vary widely, as it grows colder with ice ages and then warms again. Because we cannot be certain of the causes of those climate changes, we cannot be certain of what is causing the current warming trend.

However, we can expect, if global warming is occurring, that summers will be hotter. Over the next 100 years, sea level will rise as much as a foot or so. Is this bad? Depends upon where you live. Keep in mind, however, that not only could sea level rise 1 foot over the next 100 years, but it could continue to do so for many hundreds of years. Another point to consider is that we have routine global temperature measurements for only about 100 years. Even these are unreliable, because instruments and methods of observation changed over that course of time.

The only conclusion we can safely draw about climate and climate change is that we do not know if drastic changes are occurring. We could be at the end of a geological ice age. Evidence indicates that during interglacials, temperatures increase before they plunge. Are we ascending the peak temperature range? We have no way to tell. To what extent does our human activity impact climate? Have anthropogenic effects become so marked that we have affected the natural cycle of ice ages (which lasted for roughly the last 5 million years)? Maybe we just have a breathing spell of a few centuries before the next advance of the glaciers. If this is the case, if we are at the apogee of the current interglacial, then we have to ask ourselves a few questions: Is global warming the lesser of two evils when compared to the alternative, global cooling? If we are headed into another glacial freeze, in this era of expanding population and decreasing resources, where will we get the energy (fuel) to keep all of us warm?

ACID PRECIPITATION

Imagine an evening during a light rain—you probably feel a sense of calm and relaxation hard to describe but not hard to accept. The sound of raindrops against a roof, on the trees and lawn, on the sidewalk—all that is soothing. Whatever it is that makes you feel this way, rainfall is a major ingredient.

But someone knowledgeable and/or trained in environmental science might take another view of such a seemingly welcome and peaceful event. We might wonder to ourselves whether the rainfall is as clean and pure as it should be. Is this actually rainfall or is it rain carrying acids as strong as lemon juice or vinegar with it—capable of harming both living things like trees and lakes and nonliving things like manmade structures?

Maybe such a concern was off-the-wall before the Industrial Revolution, but today the purity of rainfall is a major concern for many people, especially its levels of acidity. Most rainfall is slightly acidic because of decomposing organic matter, the movement of the sea, and volcanic eruptions, but the principal factor is atmospheric carbon dioxide, which causes carbonic acid to form. **Acid rain** (pH <5.6) (in the pollution sense) is produced by the conversion of the primary pollutants sulfur dioxide and nitrogen oxides to sulfuric acid and nitric acid, respectively (see figure 3.2 for an explanation of pH). These processes are complex and depend on the physical

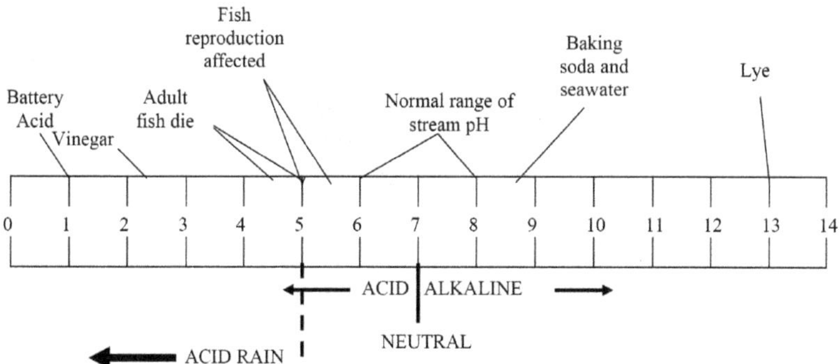

Figure 3.2. Measuring Activity: pH Scale

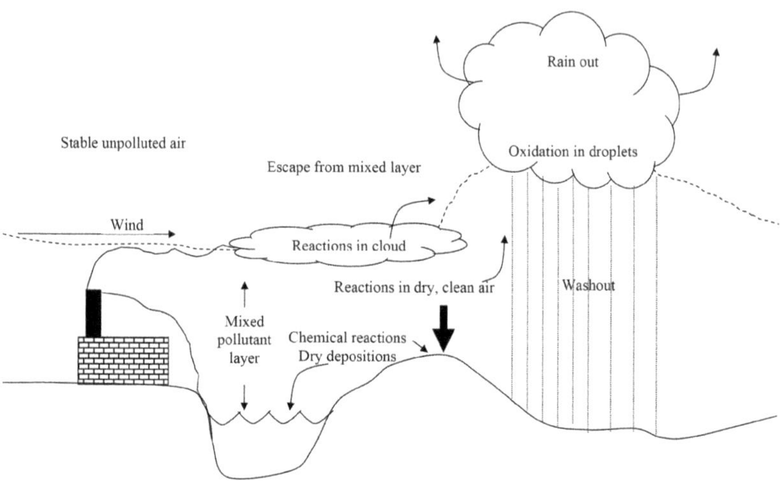

Figure 3.3. Acid Rain Cycle

dispersion processes and the rates of the chemical conversions. The basic cycle is shown in figure 3.3.

Contrary to popular belief, acid rain is not a new phenomenon, nor does it result solely from industrial pollution. Natural processes such as volcanic eruptions and forest fires produce and release acid particles into the air. The burning of forest areas to clear land in Brazil, Africa, and other countries also contributes to acid rain. However, the rise in manufacturing that began with the Industrial Revolution literally dwarfs all other contributions to the problem.

The main culprits are emissions of sulfur dioxide from the burning of fossil fuels, such as oil and coal, and nitrogen oxide, formed mostly from internal combustion engine emissions, which is readily transformed into nitrogen dioxide. These mix in the atmosphere to form sulfuric acid and nitric acid.

In dealing with atmospheric acid deposition, the earth's ecosystems are not completely defenseless; they can deal with a certain amount of acid through natural alkaline substances in soil or rocks that buffer and neutralize acid. The American Midwest and southern England are areas with highly alkaline soil (limestone and sandstone) that provide some natural neutralization. Areas with thin soil and those laid on granite bedrock, however, have little ability to neutralize acid rain.

Scientists continue to study how living beings are damaged and/or killed by acid rain. This complex subject has many variables. We know from various episodes of acid rain that the pollution can travel over very long distances. Lakes in Canada, Maine, and New York feel the effects of coal-burning in the Ohio Valley. For this and other reasons, the lakes of the world are where most of the scientific studies have taken place. In lakes, the smaller organisms often die off first, leaving the larger animals to starve to death. Sometimes the larger animals (fish) are killed directly; as lake water becomes more acidic, it dissolves **heavy metals** leading to concentrations at toxic and often lethal levels. Have you ever wandered up to the local lakeshore and observed thousands of fish belly-up? Not a pleasant sight or smell, is it? Loss of life in lakes also disrupts the system of life on the land and the air around them.

DID YOU KNOW?

While acidification of U.S. and Canadian lakes mentioned above is of major concern to environmentalists and others, it should be noted that one quarter of the carbon dioxide humans emit into the air is absorbed into the oceans. The carbon dioxide that dissolves in seawater forms carbonic acid, which in turn acidifies the ocean. The pH of the oceans has steadily dropped since the Industrial Revolution. This is an ongoing trend that is not good, especially for marine life.

In some parts of the United States, the acidity of rainfall has fallen well below 5.6. In the northeastern United States, for example, the average pH of rainfall is 4.6, and rainfalls with a pH of 4.0, which is 100 times more acidic than distilled water, are not unusual.

Despite intensive research into most aspects of acid rain, scientists still have many areas of uncertainty and disagreement. That is why the progressive, forward-thinking countries emphasize the importance of further research into acid rain.

PHOTOCHEMICAL SMOG

When various hydrocarbons, oxides of nitrogen, and sunlight come together, they can initiate a complex set of reactions that produce a number of secondary pollutants known as photochemical oxidants or photochemical smog. Smog production is a localized problem, and in regard to its impact on global climate change and global warming it is specific to that area—again, meaning that it is a problem but generally a local issue to deal with. **Photochemical smog** of the type most people are familiar with was first noticed in Los Angeles in the early 1940s. Determining its true cause has taken many years. According to Black-Covilli (1992), at first it was thought to arise from dust and smoke emitted from factories and incinerators. Accordingly, Los Angeles County officials issued a ban on all outdoor burning of trash and initiated steps toward control of industrial smoke emission. Before long, though, county authorities determined that their initial efforts were not working; the smog continued unabated. Then they went after another suspected culprit, sulfur dioxide (SO_2) given off by oil refineries and by combustion of sulfur-bearing coal. So they placed controls on sulfur dioxide emissions—but still gained no benefit. However, in the long run—for example, the reducing of vehicle emissions in Los Angeles County from 1973 to 2017—those controls have shown some beneficial improvement in lowering daily smog levels in the area.

The biochemist Dr. Arie Haagen-Smit, as a result of chance, during research aimed at finding the compounds responsible for the pleasant tastes and odors of fruit, found the cause of the smog problem, which showed beyond a doubt that the internal combustion engine was the principal source.

How do internal combustion engines produce smog? A few of the finer details are yet unclear, but the following, in simplified form, appears to be what happens. Smog begins with the high temperatures in the internal combustion engine, which cause atmospheric oxygen and nitrogen to react, producing nitric oxide ($N_2 + O_2 \rightarrow 2NO$). At the same time, varying quantities of fuel in the engine fail to burn completely. This results in a mixture of aldehydes, ketones, olefins, and aromatic hydrocarbons that is expelled in the exhaust. The exhaust enters the atmosphere, where ultraviolet radiation from the sun causes a complex series of reactions to take place. These reactions involve atmospheric oxygen, nitric oxide, and organic compounds. The result is the formation of nitrogen dioxide (NO_2) and ozone (O_3), both of which are highly toxic and irritating. In addition, this reaction also causes the formation of other constituents of photochemical smog, including formaldehyde, peroxybenzohl nitrate, peroxyacetyl nitrate (PAN), and acrolein.

Photochemical smog is known to cause many annoying respiratory effects—coughing, shortness of breath, airway constriction, headache, chest tightness, and eye, nose, and throat irritation (Masters, 1991).

STRATOSPHERIC OZONE DEPLETION

Ozone (O_3) is a molecule containing three atoms of oxygen. In the earth's stratosphere, about 50,000 to 120,000 feet high, ozone molecules band together to form a protective layer that shields the earth from some of the sun's potentially destructive ultraviolet (UV) radiation. Stratospheric ozone, formed in the atmosphere by radiation from the sun, provides us with an enormously beneficial function. Life as we know it on Earth could have evolved only with the protective ozone shield in place.

The Centers for Disease Control and Prevention in Atlanta, Georgia, look at ozone more critically, however. They point out that ozone is an extraordinarily dangerous pollutant. Only two-hundredths of a gram of ozone is a lethal dose. A single 14-ounce aerosol can filled with ozone could kill 14,000 people. Ozone is nearly as effective at destroying lung tissue as mustard gas. Not only is ozone a poisonous gas for us on Earth, it is a main contributor to air pollution, especially smog.

So, ozone is kind of a Dr. Jekyll and Mr. Hyde chemical—with a benevolent nature and an evil one. But let's look at ozone in the stratosphere.

In the early 1970s, scientists suspected that the ozone layer was being depleted. By the 1980s, it became clear that the ozone shield was indeed thinning in some places, and at times, even has a seasonal hole in it, notably over Antarctica. The exact causes and actual extent of the depletion are not yet fully known, but most scientists believe that various chemicals in the air are responsible.

Most scientists identify the family of chlorine-based compounds, most notably **chlorofluorocarbons (CFCs)** and chlorinated solvents (carbon tetrachloride and methyl chloroform), as the primary culprits involved in ozone depletion. In 1974, Molina and Rowland hypothesized the CFCs, containing chlorine, were responsible for ozone depletion. They pointed out that chlorine molecules are highly active and readily and continually break apart the three-atom ozone into the two-atom form of oxygen generally found close to Earth in the lower atmosphere.

The Interdepartmental Committee for Atmospheric Sciences (1975) estimated that a 5 percent reduction in ozone could result in nearly a 10 percent increase in cancer. This already frightening scenario was made even more frightening in 1987 when evidence showed that CFCs destroy ozone in the stratosphere above Antarctica every spring. The ozone hole had become larger, with more than half of the total ozone column wiped out and essentially all ozone having disappeared from some regions of the stratosphere (Davis and Cornwell, 1991).

In 1988, Zurer reported that on a worldwide basis, the ozone layer shrank approximately 2.5 percent in the preceding decade. This obvious thinning of the ozone layer, with its increased chances of skin cancer and cataracts, is also implicated in suppression of the human immune system and damage to other animals and plants, especially aquatic life and soybean crops. The urgency of the problem spurred the creation of the **Montreal Protocol**, an international treaty that required signatory countries to reduce their consumption of CFCs by 20 percent by 1993, and by 50 percent by 1998, and phasing out HCFCs by 2030, marking a significant achievement in solving a global environmental problem. The treaty, signed in 1987, effective in 1989, and ratified by 197 nations (and undergoing nine amendments), seems to have had a positive effect: the World Meteorological Organization (WMO) announced in a 2018 assessment that "the ozone layer [is] on the path of recovery and to potential return of the ozone values over Antarctica to pre-1980 levels by 2060." But with a very large ozone hole in Antarctica in 2020, the WMO urged, "We need to continue enforcing the Montreal Protocol banning emissions of ozone depleting chemicals" (WMO, 2020). And in other good news, the EPA reported fewer cases of skin cancer and fewer skin cancer deaths.

CLIMATE CHANGE INDICATORS

> Earth's climate is changing. Temperatures are rising, snow and rainfall patterns are shifting and more extreme climate events—like heavy rainstorms and record-high temperatures—are already taking place.
>
> —USEPA (2016)

The preceding sections discuss aspects of climate change science, global warming, and pollution, as well as the distinction between global warming and climate change. If we're going to keep a handle on climate change—both natural and human-intensified—then we need language and data to assess it. Thus we have "climate change indicators." The ones described and discussed herein are based on the official findings of the United States Environmental Protection Agency (USEPA). Note: these climate change indicators are focused on the United States.

Consider different U.S.-based cases that serve as indicators of climate change:

- In a May 13, 2019, article in *USA Today*, authors R. W. Mill and D. Rice point out that data from the Mauna Loa Observatory in Hawaii showed that carbon dioxide levels surpassed 415 parts per million (ppm). This is the highest reading recorded since humans existed millions of years ago. The point is that even accounting for natural fluctuations historical measurements show that the current global atmospheric concentrations of carbon dioxide are unprecedented compared with the past 800,000 years (USEPA, 2016).
- As sea level rises flooding is becoming more frequent along the U.S. coastline. USEPA (2016) points out that nearly every site measured has experienced an increase in coastal flooding since the 1950s. Along many locations along the East and Gulf coasts the rate of flooding is accelerating.
- In the measurement sites in the western United States between 1955 and 2016 snowpack in early spring has decreased more than 90 percent.
- Another issue is marine species distribution whereby the average center of biomass for more than 100 marine fish and invertebrate species along U.S. coasts has shifted northward approximately 10 miles between 1982 and 2015. Moreover, many of these species have moved an average of 20 feet deeper (USEPA, 2016).

HOW MANY INDICATORS ARE THERE?

Currently, USEPA presents 37 indicators of a specific climate-change topic, such as Greenhouse Gas Emissions. In the following chapters we present six major indicator categories with specific descriptions of sub-indicators within each major category. These major indicator categories, as listed and detailed by USEPA, are:

- Greenhouse Gases
- Weather and Climate
- Oceans
- Snow and Ice
- Health and Society
- Ecosystems

The reader may wonder why it is important to use indicators in relation to climate change. Indicators are used simply because they are an important way to track and communicate the causes and effects of climate change. Keep in mind that an indicator represents the state or trend of certain environmental or societal conditions over a given area and a specified period of time. Another key point is that all the indicators discussed in this book correlate to either the causes or effects of climate change. For those who directly connect human actions involved with climate change, the indicators show trends that actually can more directly be linked to human-induced climate change than others. Collectively, the trends depicted in these indicators provide important evidence of "what climate change looks like" (USEPA, 2016).

REFERENCES AND RECOMMENDED READING

Black-Covilli, L. L. 1992. Basic Air Quality, in *Fundamentals of Environmental Science and Technology*. Porter-C. Knowles, ed. Rockville, MD: Government Institutes.

Broecker, W. 1987. Unpleasant Surprises in the Greenhouse? *Nature* 328: 123–26.

Bruun, P. 1962. Sea Level Rise as a Cause of Shore Erosion. *Proceedings of the American Society of Engineers and Journal Waterways Harbors Division* 88: 117–30.

Bruun, P. 1986. Worldwide Impacts of Sea Level Rise on Shorelines. *Effects of Changes in Stratospheric Ozone and Global Climate*, vol. 4, New York: UNEP/EPA, pp. 99–128.

Crowley, T. J., and G. R. North. 1996. Abrupt Climate Change and Extinction Events in Earth's History. *Science* 240.

Davis, M. L., and D. A. Cornwell. 1991. *Introduction to Environmental Engineering*. New York: McGraw-Hill.

Dolan, E. F. 1991. *Our Poisoned Sky*. New York: Cobblehill Book.

Edgerton, L. 1991. *The Rising Tide: Global Warming and World Sea Levels*. Washington, DC: Island Press.

Franck, I., and D. Brownstone. 1992. *The Green Encyclopedia*. New York: Prentice-Hall.

Graedel, T. E., and P. J. Crutzen. 1989. The Changing Atmosphere. *Scientific American* (September): 58–68.

Hansen, J. E., et al. 1986. Climate Sensitivity to Increasing Greenhouses Gases. *Greenhouse Effect and Sea Level Rise: A Challenge for This Generation*. M. C. Barth & J. G. Titus, ed. New York: Van Nostrand Reinhold.

Hansen, J. E., et al. 1989. Greenhouse Effect of Chlorofluorocarbons and Other Trace Gases. *Journal of Geophysical Research* 94 (November): 16417–21.

Hegerl, G. C., F. W. Zwiers, P. Braconnot et al. 2007. Chapter 9: Understand and Attributing Climate Change, Section 9.2.2, Spatial and Temporal Patterns of the Response to Different Forcings and Their Uncertainties, in S. Solomon, D. Qin, M. Manning et al., *Climate Change 2007: The Physical Science Basis. Contribution of Working Group I to the Fourth Assessment Report of the Intergovernmental Panel on Climate Change*. Intergovernmental Panel on Climate Change. Cambridge, UK: Cambridge University Press.

Interdependent Committee for Atmospheric Sciences. 1975. *The Possible Impact of Fluorocarbons and Hydrocarbons on Ozone*. Washington, DC: U.S. Government Printing Office, p. 3, May.

Masters, G. M. 1991. *Introduction to Environmental Engineering and Science*. Englewood Cliffs, NJ: Prentice-Hall.

Molina, M. J., and F. S. Rowland. 1974. Stratospheric Sink for Chlorofluoromethanes: Chlorine Atom Catalyzed Destruction of Ozone. *Nature* 248: 810–12.

NASA. 2019. Global Climate Change: Vital Signs of the Planet. Accessed 3/2/19 @ https://climate.nasa.gov/resources/global-warming.

Nature. 2002. Contrails Reduce Daily Temperature Range. August 8. Accessed March 2, 2009 @ www.nature.com/nature.

Ramanathan, V. 2006. Atmospheric Brown Clouds: Health, Climate and Agriculture Impacts. *Pontifical Academy of Sciences Scripta Varia* 106: 47–60.

Spellman, F. R., and N. Whiting. 2006. *Environmental Science and Technology: Concepts and Applications*. Boca Raton, FL: CRC Press.

Stanhill, G., and S. Moreshet. 2004. Global Radiation Climate Changes in Israel. *Climatic Change* 22: 121–38.

Thompson, D. J. 1995. The Seasons, Global Temperature, and Precession. *Science* 268: 59.

Time. 1998. Global Warming: It's Here . . . And Almost Certain to Get Worse. August 24.

USA TODAY. 1997. Global Warming: Politics and Economics Further Complicate the Issue. December 1, pp. A-1, 2.

USA TODAY. 2009. Your Eyes Aren't Deceiving You: Skies Are Dimmer. March 12.

USEPA. 2005. *Basic Air Pollution Meteorology*. Accessed 01/15/08 @ www.epa.gov/apti.

USEPA. 2007. *National Ambient Air Quality Standards* (*NAAQS*). Accessed January 12, 2008 @ www.epa.gov/air/criteria/html.

USEPA. 2009. *Regulatory Atmospheric Modeling*. Accessed March 2, 2009 @ http://www.epa.gov/scram001.

USEPA. 2010. *Climate Change and Ecosystems*. Accessed June 2, 2011 @ www.epa.gov/climate change.

USEPA. 2016. Climate Change Indicators in the United Sates, 2016. Fourth edition. EPA-430-R-16-004. http://www.epa.gov/climate-indicators.

USEPA. 2017. *The Probability of Sea Level Rise*, Washington, DC: Environmental Protection Agency.

USGS. 2009. *Hydroelectric Power Water Use*. Accessed November 5, 2009 @ http://ga.water.usgs.bof/edu/suhy.html.

Wigley, T. M., P. D. Jones, and P.M. Kelly. Empirical Climate Studies: Warm World Scenarios and the Detection of Climatic Change Induced by Radioactively Active Gases. *The Greenhouse Effect, Climatic Change, and Ecosystems*. B. Bolin et al., ed. New York: Wiley, 1986.

WMO. 2009. *Manual of Codes*. Accessed March 2, 2009 @ http://www.wmo.ch/pages/prog/www/WMOCodes/Manual/WMO306_vol-1-2-PartB.pdf.

WMO. 2020. 2020 Antarctic Ozone Hole Is Large and Deep. https://public.wmo.int/en/media/news/2020-antarctic-ozone-hole-large-and-deep.

Wood, R., et al. 2007. Climate Models and Their Evaluation. In S. Solomon, D. Qin, M. Manning et al. *Climate Change 2007: The Physical Science Basins*. Cambridge, UK: Cambridge University Press.

Zurer, P. S. 1988. Studies on Ozone Destruction Expand Beyond Antarctic. *C & E News*, May, pp. 18–25.

Chapter 4

Climate Change Indicator One

Greenhouse Gases

Mexico Cancels School in Capital Due to Poor Air Quality

—Associated Press headline, May 2019

Greenhouse gases (GHGs) warm the earth by absorbing energy and slowing the rate at which the energy escapes to space; they act like a blanket insulating the planet. Different GHGs can have different effects on the earth's warming. Two key ways in which these gases differ from each other are their ability to absorb energy (their "radiative efficiency"), and how long they stay in the atmosphere (also known as their "lifetime") (USEPA, 2017).

We mentioned greenhouse gases several times in the introductory sections of this book. In this chapter we go deep—real deep into the subject matter. Greenhouse gases that result from human activities are the most significant driver of observed climate change since the mid-twentieth century (NOAA, 2016). So indicator number one focuses on the emissions of the major greenhouse gases resulting from human activities, the concentrations of these gases in the atmosphere, and how emissions and concentrations have changed over time. Note that when comparing gases, these indicators use a concept called "global warming potential" (GWP) to convert amounts of other gases into carbon dioxide equivalents.

WHAT ARE GLOBAL WARMING POTENTIALS?

The Global Warming Potential (GWP) was developed to allow comparisons of the global warming impacts of different gases. GWP is a measure how much energy the emissions of 1 ton of a gas will absorb over a given period of time, relative to the emissions of 1 ton of carbon dioxide (CO_2). The larger the GWP, the more that a given gas warms the earth compared to CO_2 over that time period. The time period usually used for GWPs is 100 years. GWPs provide a common unit of measure, which allows analysts to add up emissions estimates of different gases (e.g., to compile a national GHG inventory), and allows policy makers to compare emissions reduction opportunities across sectors and gases.

- Carbon dioxide (CO_2), by definition, has a GWP of 1 regardless of the time period used, because it is the gas being used as the reference. CO_2 remains in the climate system for a very long time: CO_2 emissions cause increases in atmospheric concentration of CO_2 that will last thousands of years.
- Methane (CH_4) is estimated to have a GWP of 28–36 over 100 years. CH_4 emitted today lasts about a decade on average, which is much less than CO_2. But CH_4 also absorbs much more energy than CO_2. The net effect of the shorter lifetime and higher energy absorption is reflected in the GWP. CH_4 is also a precursor to ozone, and ozone itself is a GHG.
- Nitrous oxide (N_2O) has a GWP 265–298 times that of CO_2 for a 100-year timescale. N_2O emitted today remains in the atmosphere for more than 100 years, on average.
- High-GWP gases such as chlorofluorocarbons (CFCs), hydrofluorocarbons (HFCs), hydrochlorofluorocarbons (HCFCs), perfluorocarbons (PFCs), and sulfur hexafluoride (SF_6), for a given amount of mass, trap substantially more heat than CO_2 and their GWPs can be in the thousands or tens of thousands (USEPA, 2017). Table 4.1 lists GWPs based on 2007 data from IPCC (2007).

Table 4.1. Global Warming Potentials (100-Year Time Horizon)

Gas	*GWP*
CO_2 (Reference Gas)	1
CH_4*	25
N_2O	298
HFC-23	14,800
HFC-32	675
HFC-125	3,500
HFC-134a	1,430
HFC-143a	4,470
HFC-152a	124
HFC-227ea	3,220
HFC-236fa	9,810
HFC-4310mee	1,640
CF_4	7,390
C_2F_6	12,200
C_4F_{10}	8,860
C_6F_{14}	9,300
SF_6	22,800
NF_3	17,200

*Includes direct effects due to tropospheric ozone and stratospheric water vapor.

Source: IPCC (2007).

U.S. GREENHOUSE GAS EMISSIONS

A number of activities influence the magnitude of greenhouse gases released into the atmosphere, including economic activity, population, consumption patterns, energy prices, land use, and technology. In tracking these emissions in the United States we have a number of ways to do this, such as by measuring emissions directly, calculating emissions based on the amount of fuel that people burn, and estimating other activities and their associated emissions.

So, the question becomes what is going on or happening at the present time or recently in regard to U.S. greenhouse gas emission?

Well, let's take a look at a few of the happenings going with U.S. greenhouse gas emissions. In 2014, for example, U.S. greenhouse gas emissions totaled 6,870 million metric tons (15.1 trillion pounds) of carbon dioxide equivalents. This 2014 total represents a 7 percent increase since 1990 but a 7 percent decrease since 2005. It is important to point out that among the various sectors of the U.S. economy, electricity generation (power plants) accounts for the largest share of emissions—31 percent of total greenhouse gas emissions since 1990. Transportation is the second-largest sector, accounting for 26 percent of the emissions since the 1990s. It is interesting to note that emissions sinks, the opposite of emissions sources, absorb carbon dioxide from the atmosphere. In 2014, 11 percent of U.S. greenhouse gas emissions were offset by **net sinks** resulting from land use and forestry practices. Growing forests remove carbon form the atmosphere, outweighing emission from wildfires. Other carbon emission and sinks result from crop practices, burning biofuels, or depositing yard trimmings and food scraps in landfills (USEPA, 2016).

RECENT TRENDS IN U.S. GREENHOUSE GAS EMISSIONS AND SINKS

It is interesting to note that in 2017, total gross U.S. greenhouse gas emissions were 6,456 million metric tons (MMT) of carbon dioxide (CO_2). Total U.S. emissions have increased by 1.3 percent from 1990 to 2017, and emissions decreased from 2016 to 2017 by 0.5 percent (33.5 MMT CO_2). The decrease in total greenhouse gas emissions between 2016 and 2017 was driven in part by a decrease in CO_2 emissions from fossil fuel combustion. The decrease in CO_2 emissions from fossil fuel combustion was a result of multiple factors, including a continued shift from coal to natural gas and increased use of renewable energy in the electric power sector, and milder weather that contributed to less overall electricity use (USEPA, 2017).

GLOBAL GREENHOUSE GAS EMISSIONS

Substantial increase in atmospheric concentrations of greenhouse gases due to human activities has occurred and is continuing. Every country around the world emits greenhouse gases into the atmosphere, meaning the root cause of climate change is truly global in scope. Some countries produce far more greenhouse gases than others, and several factors—such as economic activity, population, income level, land use, and climate conditions—can influence a country's emissions levels. Tracking greenhouse gas emissions worldwide provides a global context for understanding the United States' and other nations' roles in climate change.

So, what exactly is happening with regard to increases in atmospheric concentrations of greenhouse gases? In the first place we need to look at 2010. In 2010, estimated worldwide emissions from human activities totaled nearly 46 billion metric tons of greenhouse gases, expressed as carbon dioxide equivalents. This represents a 35 percent increase from 1990. These numbers represent net emissions, which include the effects of land use and forestry.

Between 1990 and 2010, global emissions of all major greenhouse gases increased. Net emissions of carbon dioxide increased by 42 percent, which is particularly important because carbon dioxide accounts for about three-fourths of total global emissions. Nitrous oxide emissions increased the least—9 percent—while emission of methane increased by 15 percent. Emissions of fluorinated gases more than doubled (USEPA, 2017).

ATMOSPHERIC CONCENTRATIONS OF GREENHOUSE GASES

Since the mid-1700s when the Industrial Revolution began, people have added a substantial amount of heat-trapping greenhouse gases into the atmosphere by burning fossil fuels, cutting down forests, and conducting other activities. Many of these gases remain in the atmosphere for long time periods ranging from a decade to many millennia, which allows them to become well mixed throughout the global atmosphere. As a result of human activities, these gases are entering the atmosphere more quickly than they are being removed by chemical reactions or by emissions sinks, such as the oceans and vegetation, which absorb greenhouse gases from the atmosphere. Thus, their concentrations are increasing, which contributes to global warming (USEPA, 2017).

When we look at what's happening with atmospheric concentrations of greenhouse gases a couple of significant aspects is apparent. First, global atmospheric concentrations of carbon dioxide have risen significantly over the last few hundred years. Secondly, historical measurements show that the current global atmospheric concentrations of carbon dioxide are unprecedented compared with the past 800,000 years. Another occurrence happening at the present is since the beginning of the industrial era, concentrations of carbon dioxide have increased from an annual average of 280 ppm in the late 1700s to 401 ppm measured at Mauna Loa in 2015—a 43 percent increase due to human activities (IPCC, 2013).

DID YOU KNOW?

PPM (parts per million) may be roughly described as an amount contained in a full shot glass in the bottom of a standard-sized swimming pool—the full shot glass is 1 ppm relative to the rest of the water contained in the pool.

CLIMATE FORCING

The Intergovernmental Panel on Climate Change (IPCC) defines **climate forcing** as "an externally imposed perturbation in the radiative energy budget of the earth climate system, for example, through changes in solar radiation, changes in the earth's albedo or changes in atmospheric gases and aerosol particles." Thus climate forcing is a "change" in the status quo. IPCC takes the preindustrial era (before 1750) as the baseline. The perturbation to direct climate forcing (also termed "radiative forcing") that has the largest magnitude and the least scientific uncertainty is the forcing related to changes in long-lived, well mixed greenhouse gases, in particular carbon dioxide (CO_2), methane (CH_4), nitrous oxide (N_2O), and halogenated compounds (mainly CFCs).

This "radiative forcing" or heating effect is continually measured (see figure 4.1). Radiative forcing occurs when energy from the sun reaches the earth and the planet absorbs some of this energy and radiates the rest back as heat. Note that a variety of physical and chemical factors—some natural and some influenced by humans—can shift the balance between incoming and outgoing energy, which forces changes to the earth's climate. These changes are measured by the amount of warming or cooling they can produce. When the changes have a warming effect they are called "positive" forcing, whereas those that have a cooling effect are called "negative" forcing. When positive and negative forces are

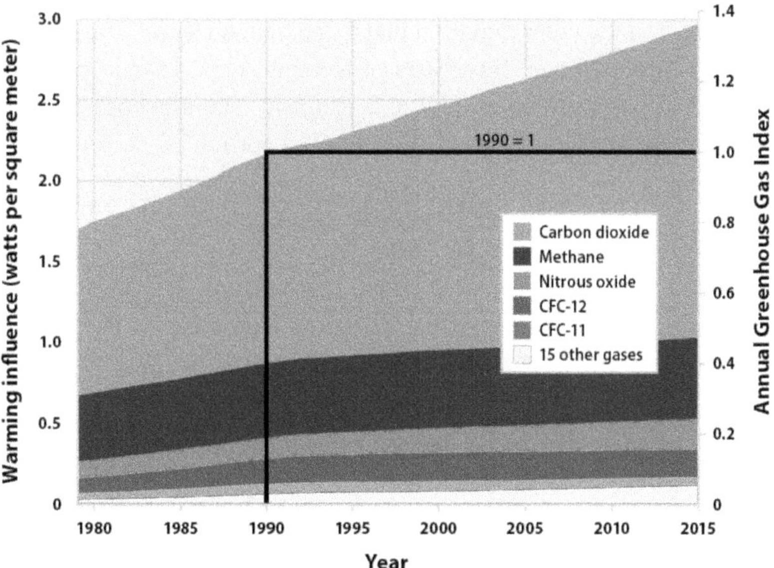

Figure 4.1. Radiative Forcing

out of balance the result is a change in the earth's average surface temperature. Greenhouse gases trap heat in the lower atmosphere and cause positive radiative forcing (NOAA, 2018).

Atmospheric global greenhouse gas abundances are used to calculate changes in radiative forcing beginning in 1979 when NOAA's global air sampling network expanded significantly. NOAA's Annual Greenhouse Gas Index is defined by the change in annual average total radiative forcing by all the long-lived greenhouse gases since the preindustrial era. The Annual Greenhouse Gas Index was introduced in 2006 and is based on measurements through 2004 (Hofmann et al., 2008).

WHAT IS ALBEDO?

With regard to climate forcing or radiative forcing **albedo** (the ratio of light reflected [reflectivity] from a particle, planet, or satellite to that falling on it) always has a value less than or equal to 1—an object with a high albedo (near 1) is very bright, while a body with a low albedo (near 0) is dark. For example, freshly fallen snow typically has an albedo that is between 75 and 90+ percent; that is, 75 to 95 percent of the solar radiation that is incident on snow is reflected. At the other extreme, the albedo of a rough, dark surface, such as a green forest, may be as low as 5 percent. The albedos of some common surfaces are listed in table 4.2 The portion of insolation not

Table 4.2. Albedo of Surface Types (percent reflected)

Surface	*Albedo (%)*
Water (low sun)	10–100
Water (high sun)	3–10
Grass	16–26
Glacier ice	20–40
Deciduous forest	15–20
Coniferous forest	5–15
Old snow	40–70
Fresh snow	75–95
Sea ice	30–40
Blacktopped tarmac	5–10
Desert	25–30
Crops	15–25

reflected is absorbed by the earth's surface, warming it. This means Earth's albedo plays an important part in the earth's radiation balance and influences the **mean annual temperature** and the **climate**, on both local and global scales.

NOAA MONITORING PROGRAM

Radiative climate forcing measurements of the global abundance and distribution of long-lived greenhouse gases are used to calculate changes in radiative climate forcing. NOAA (National Oceanic and Atmospheric Administration) collects air samples, which include cooperative programs for the carbon gases, providing samples from approximately 80 global background air sites, including measurements at 5-degree latitude intervals from ship routes.

A global average is calculated from a smoothed north-south profile using weekly data (see figure 4.2). The atmospheric abundance

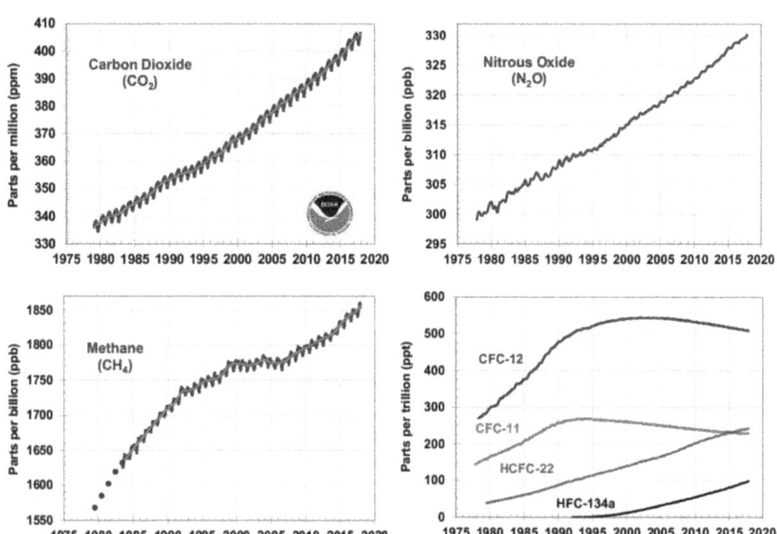

Figure 4.2. Global Average Abundances of the Major, Well-Mixed, Long-Lived Greenhouse Gases

Note: Carbon dioxide, methane, nitrous oxide, CFC-12, and CFC-11 from the NOAA global air-sampling network are plotted since the beginning of 1979. These five gases account for about 96 percent of the direct radiative forcing by long-lived greenhouse gases since 1750. The remaining 4 percent is contributed by an assortment of 15 minor halogenated gases including HCFC-22 and HFC-134a, for which NOAA observations are also shown.

Sources: Methane data before 1983 are annual averages obtained by NOAA from D. Etheridge et al. (1998) and then adjusted to the NOASS calibration scale (Dlugokencky et al., 2005; NOAA, 2018).

of CO_2 has increased by an average of more than 1.81 ppm per year over the past 41 years (1979–2019). Note that the CO_2 increase is accelerating—while it averaged about 1.6 ppm per year in the 1980s and 1.5 ppm per year in the 1990s, the growth rate increased to 2.2 ppm per year during the last decade (2008–17). The annual CO_2 increase from 1 Jan 2017 to 1 Jan 2018 was 2.3 ± 0.1 ppm (NOAA, 2018).

A decline in the growth rate of methane was recorded from 1983 until 1999. Superimposed on this decline is significant interannual variability in growth rates (Dlugokencky et al. 2005). From 1999 to 2006, the atmospheric CH_4 burden was nearly constant, but since 2007, globally averaged CH_4 has been increasing again. Causes for the increase during 2007–8 included warm temperatures in the Arctic in 2007 and increased precipitation in the tropics during 2007 and 2008 (Dlugokencky et al., 2009). Continued increasing microbial emissions after 2008 likely from wetlands or agriculture seem certain based on isotopic measurements (Schaefer et al., 2016; Nisbet et al., 2016). Some recent papers have also suggested contribution to the plateau and subsequent increase in methane's global abundance from changes in the loss rate of methane (Rigby et al., 2017). Since 2013, the global within-year increase (January 1 to December 31) methane has become even larger, with increases between 8.8 ± 2.6 through 2017 compared to an average annual increase of 5.7 ± 1.1 ppb yr between 2007 and 2013. The atmospheric burden of nitrous oxide continues to slowly increase over time, with an average rate of 0.9 ppb yr over the past decade. Radiative forcing from the sum of observed CFC changes ceased increasing in about 2000 and continued to decline through 2017 (Montzka et al., 2011).

If we were to fashion an index based on the total of these contributions to radiative forcing it would be similar to the Consumer Price Index, for example. However, even with all the important components included not all the components of climate forcing would be present. The results reported here are based mainly on measurements of long-lived, well-mixed gases and have small uncertainties (NOAA, 2018).

MAJOR AND MINOR RADIATIVE FORCING GASES

Unsurprisingly, CO_2 dominates the total forcing with methane and the CFCs becoming relatively smaller contributors to the total forcing over time. The five major greenhouse gases account for about 96 percent of the direct radiative forcing by long-lived greenhouse gas increases since 1750. The 15 minor halogenated gases (CFC-113, CCl_4, CH_3CCL_3, HCFCs 22, 141b, and 142b, HFCs 134a, 152a, 23, 143a, and 125, SF_6, and halons 1211, 1301, and 2402) contribute the remaining 4 percent.

Of the five long-lived greenhouse gases, CO_2 and N_2O are the only ones that continue to increase at regular rates over decades. Radiative forcing from CH_4 increased since 2007 after remaining nearly constant from 1999 to 2006. While the radiative forcing of the long-lived, well-mixed greenhouse gases increased 41 percent from 1990

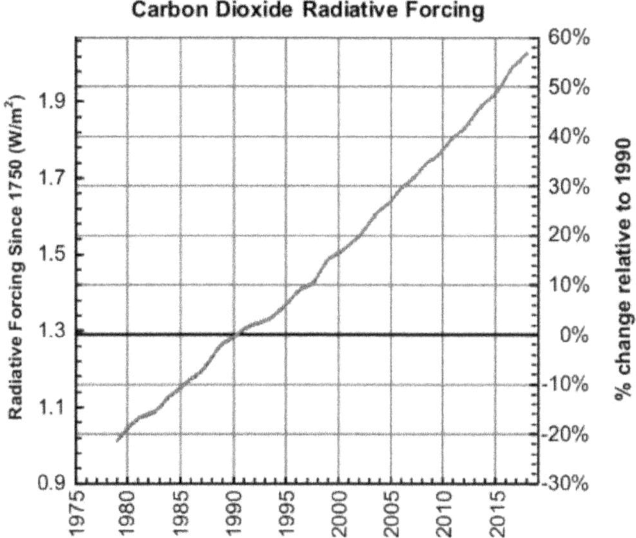

Figure 4.3. Relative Forcing, Relative to 1750, due to Carbon Dioxide Alone since 1979

Note: The recent change from January 1, 1990, is shown on the right axis and shows how the direct warming influence of CO_2 on climate has increased by about 50 percent since 1990.

Source: NOAA (2018).

COWS AND METHANE

Let's take a look at cows and their production of methane. The source of cow methane is not primarily from flatulence, contrary to popular belief. Instead, the bigger methane source is from cow belching. Cows—like all ruminants—belch because of enteric fermentation—the digestive process of converting sugars into simple molecules for absorption into the bloodstream, which produces methane as a by-product. The methane produced by the cow that is the result of flatulence (farting) comes from the large intestine and is expelled from the cow's rear end.

In the United States, it is the burning of fossil fuels that produces most of the greenhouse gases. About 2 percent of the *methane* produced comes from ruminants such as cows (USEPA, 2018). Currently, there are 95 million cattle in the United States (Fountain, 2020). Globally, wetlands (bogs, etc.) are responsible for 30 percent of methane emissions, followed by agriculture (which would include manure management) at 24 percent, and burning of fossil fuels at 20 percent (Ivanov, 2019).

Interestingly, a study published in spring of 2021 found that when red seaweed (*Asparagopsis taxiformis*) was given to beef steers as a feed additive, the production of methane dropped by 80 percent. In some study groups, the reduction was 99 percent (Roque et al., 2021). The researchers found no reduction in beef productivity or meat quality.

to 2017, CO_2 has accounted for about 80 percent of amendments, and it is estimated that climate forcing would have been as much as 0.3 watt m^{-2} greater in 2010 (Velders et al., 2007), or more than half of the increase in radiative forcing due to CO_2 alone since 1990.

REFERENCES AND RECOMMENDED READING

Associated Press. 2019. Mexico Cancels School in Capital Due to Poor Air Quality. May 15, 2019. https://apnews.com/article/8225b6b4f2354c1c9e7c7dfd67196e92.

Dlugokencky, E. J., R. C. Myers, P. M. Lang et al. 2005. Conversion of NOAA Atmospheric Dry Air CH_4 Mole Fractions to a Gravimetrically Prepared Standard Scale. *J. Geophys. Res.* 110: D18306, doi:10.1029/2005JD006035.

Dlugokencky, E. J., L. Bruhwiler, J. W. C. White, et al. 2009. Observational Constraints on Recent Increases in CH4 Burden in the Atmosphere. *Geophys. Res. Lett. 36*(18).

Etheridge, D. M., L. P. Steele, R. J. Francey, and R. L. Langenfelds. 1998. Atmospheric Methane between 1000 A.D. and Present: Evidence of Anthropogenic Emissions and Climate Variability. *J. Geophys. Res.* 103: 15979–993.

Fountain, Henry. 2020. Belching Cows and Endless Feedlots: Fixing Cattle's Climate Issues. *New York Times*. October 21.

Hofmann, D. J., J. H. Butler, E. J. Dlugokencky et al. 2008. The Role of Carbon Dioxide in Climate Forcing from 1979–2004: Introduction of the Annual Greenhouse Gas Index. *Tellus B* 58B: 614–19.

IPCC. 2007. *Climate Change 2007: The Physical Science Basis. Contribution of Working Group 1 the Fourth Assessment Report of the Intergovernmental Panel on Climate Change.* S. Solomon, D. Qin, M. Manning et al., eds. Cambridge, UK: Cambridge University Press.

IPCC. 2013. Climate Change 2013: The Physical Science Basis. Working Group 1 contribution to the IPCC Fifth Assessment Report. Cambridge, UK: Cambridge University Press. www.ipcc.ch/report/ar5/wg1.

Ivanova, Irina. 2019. Who Are the Biggest U.S. Methane Emitters? *CBS News*. Accessed February 2021 @ https://www.cbsnews.com/news/who-are-the-biggest-us-methane-emitters.

Montzka, S. A., E. J. Dlugokencky, and J. H. Butler. 2011. Non-CO_2 Greenhouse Gases and Climate Change. *Nature* 476: 43–50.

Nisbet, E. G., et al. 2016. Rising Atmospheric Methane: 2007–2014 Growth and Isotopic Shift. *Global Biogeochem, Cycles* 30.

NOAA. 2016. The NOAA Annual Greenhouse Gas Index. Accessed May 18, 2019 @ https://www.esrl.noaa.gov/gmd/aggi/aggi.html.

Rigby, M., et al. 2017. Role of Atmospheric Oxidation in Recent Methane Growth. *Proc. Natl. Acad. Sci.* https://doi.org/10.1073/pnas.1616426114.

Roque, Breanna M., M. Ermias Kebreab, R. D. Venegas et al. 2021. Red Seaweed (*Asparagopsis taxiformis*) Supplementation Reduces Enteric Methane by over 80 Percent in Beef Steers. Accessed March 2021 @ https://journals.plos.org/plosone/article?id=10.1371/journal.pone.0247820.

Schaefer, H., et al. 2016. A 21st Century Shift from Fossil-Fuel to Biogenic Methane Emissions Indicated by CH_4. *Science* 352: 80–84.

USEPA. 2016. Climate Change Indicators in the United States, 2016. Fourth edition. EPA-430-R-16-004. www.epa.gov/climate-indicators.

USEPA. 2017. Understanding Global Warming Potentials. Accessed May 16, 2019 @ https://www.epa.gov/ghgemissions/understanding-global-warming-potentials.

USEPA. 2018. Sources of Greenhouse Gas Emissions: Agriculture Sector Emissions. Accessed February 2021 @ https://www.epa.gov/ghgemissions/sources-greenhouse-gas-emissions#agriculture.

Velders, G. J. M., S. O. Andersen, J. S. Daniel, D. W. Fahey, and M. McFarland. 2007. The Importance of the Montreal Protocol in Protecting Climate. *Proc. Nat. Acad. Sciences* 104: 4814–19.

Chapter 5

Climate Change Indicator Two

Weather and Climate

> Moistening, misting, and heaving the earth breathes like a great dark beast. When barometric pressure is high, the earth holds its breath and vapors lodge in the loose packing and random crannies of the soil, only to float out again when the pressure is low and the earth exhales. The keen-nosed, like Helen Keller, smell the vapors rising from the soil, and know by that signal that there will be rain or snow. This may also be, in part, how farm animals anticipate earthquakes—by smelling ions escaped from the earth.
>
> —Diane Ackerman, from *A Natural History of the Senses*

WEATHER VS. CLIMATE[1]

Have you ever asked someone how their climate is—when you really wanted to know what their weather was? We don't often confuse the two. When we talk about **weather**, we are generally referring to the transient changes in temperature, precipitation, and wind that affect whether we take the umbrella along or wear a heavy coat. Most people rely heavily on the local meteorologist and the daily weather forecasts: so much so that an entire (and very visible) branch of science is dedicated to the effort of trying

1. Adapted from NOAA's National Weather Service Public Affairs Office. Evolution of the National Weather Service (2010). Accessed 05/20/19 @ http://www.nws.noaa.gov/pa/history/timeline.php.

to predict the weather—a difficult task because of the extensive variables in any prediction.

Weather is the state of the atmosphere, mainly with respect to its effect upon life and human activities. As distinguished from **climate** (long-term manifestations of weather), weather consists of the short-term (minutes to months) variations of the atmosphere. Weather is defined primarily in terms of heat (temperature), pressure, clouds, humidity, wind, and moisture. At high levels above the earth, where the atmosphere thins to near vacuum, weather does not exist. Weather is a near-surface phenomenon. We see this clearly, daily, as we observe the ever-changing, sometimes dramatic, and often violent weather display that travels through our environment.

And while we're looking at definitions, **meteorology** is the science concerned with the atmosphere and its phenomena; the meteorologist observes the atmosphere's temperature, density, winds, clouds, precipitation, and other characteristics, and endeavors to account for its observed structure and evaluation (weather, in part) in terms of external influence and the basic laws of physics. Meteorological phenomena affect the chemical properties of the **atmosphere**—the gaseous envelope surrounding Earth.

In the study of air quality (especially regarding air pollution in a particular area), the determining factors are directly related to the dynamics of the atmosphere—local weather. These determining factors include strength of winds, the direction they are blowing, temperature, available sunlight (needed to trigger photochemical reactions, which produce smog), and the length of time since the last weather event (strong winds and heavy precipitation) cleared the air.

WINDS AND BREEZES

On bright clear nights, the earth cools more rapidly than on cloudy nights, because cloud cover reflects a large amount of heat back to Earth, where it is reabsorbed again. Air is heated primarily by contact with the warm earth. When air is warmed, it expands and becomes lighter. Air warmed by contact with the earth rises and is replaced by cold air that flows in and under it. When this cold air is warmed, it, too, rises, and is replaced by cold air. This cycle continues and generates a circulation of warm and cold air called **convection**.

At the earth's equator, the air receives much more heat than the air at the poles. This warm air at the equator is replaced by colder air flowing in from north and south. The warm, light air rises and moves poleward high above the earth. As it cools, it sinks, replacing the cool surface air that has moved toward the equator.

It is the circulating movement of warm and cold air (convection) and the differences in heating that cause local **winds** and **breezes**. Different amounts of heat are absorbed by different land and water surfaces. Soil that is dark and freshly plowed absorbs much more than grassy fields, for example. Land warms faster than water during the day and cools faster at night. Consequently, the air above such surfaces is warmed and cooled, resulting in production of local winds.

Winds should not be confused with air currents. Wind is primarily oriented toward horizontal flow, while **air currents** are created by air moving upward and downward. Both wind and air currents have direct impact on air pollution, which is carried and dispersed by wind. An important factor in determining the locations most affected by an air pollution source is wind direction. Since air pollution is a global problem, wind direction on a global scale is important (see figure 5.1).

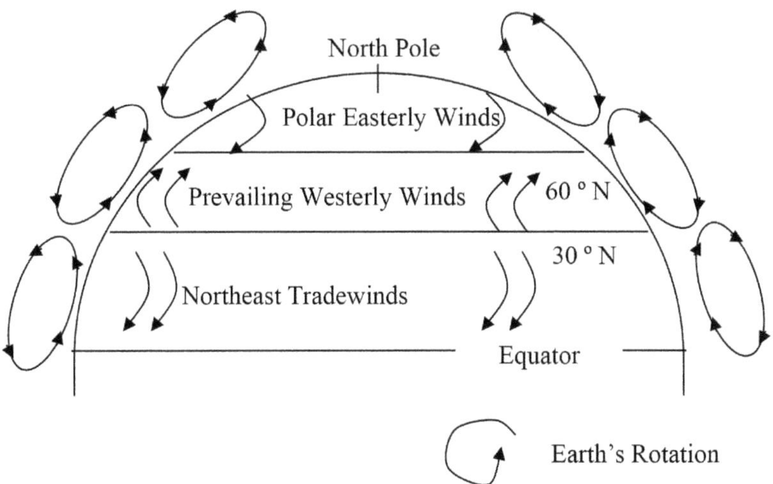

Figure 5.1. Wind Directions in the Northern Hemisphere

AIR MASSES

An **air mass** is a vast body of air (so vast that it can have global implications) in which the condition of temperature and moisture are much the same at all points in a horizontal direction. An air mass is affected by and takes on the temperature and moisture characteristics of the surface over which it forms, though its original characteristics tend to persist.

When two different air masses collide, a **front** is formed. A **cold front** marks the line of advance of a cold air mass from below as it displaces a warm air mass. A **warm front** marks the advance of a warm air mass as it rises up over a cold one.

THERMAL INVERSIONS AND AIR POLLUTION

We've said that during the day the sun warms the air near the earth's surface. Normally, this heated air expands and rises during the day, diluting low-lying pollutants and carrying them higher into the atmosphere. Air from surrounding high-pressure areas then moves down into the low-pressure area created when the hot air rises. This continual mixing of the air helps keep pollutants from reaching dangerous levels in the air near the ground.

Sometimes, however, a layer of dense, cool air is trapped beneath a layer of less dense, warm air in a valley or urban basin. This is called a **thermal inversion** (see figure 5.2). In effect, a warm-air lid covers the region and prevents pollutants from escaping in upward-flowing air currents. Usually these inversions trap air pollutants at ground level for a short period of time. However, sometimes they last for several days, when a high-pressure air mass stalls over an area, trapping air pollutants at ground level where they accumulate to dangerous levels.

The best-known location in the United States where thermal inversions occur almost on a daily basis is in the Los Angeles basin. The Los Angeles basin is a valley with a warm climate, light winds, surrounded by mountains located near the Pacific Coast. Los Angeles itself is a large city with a large population of people and

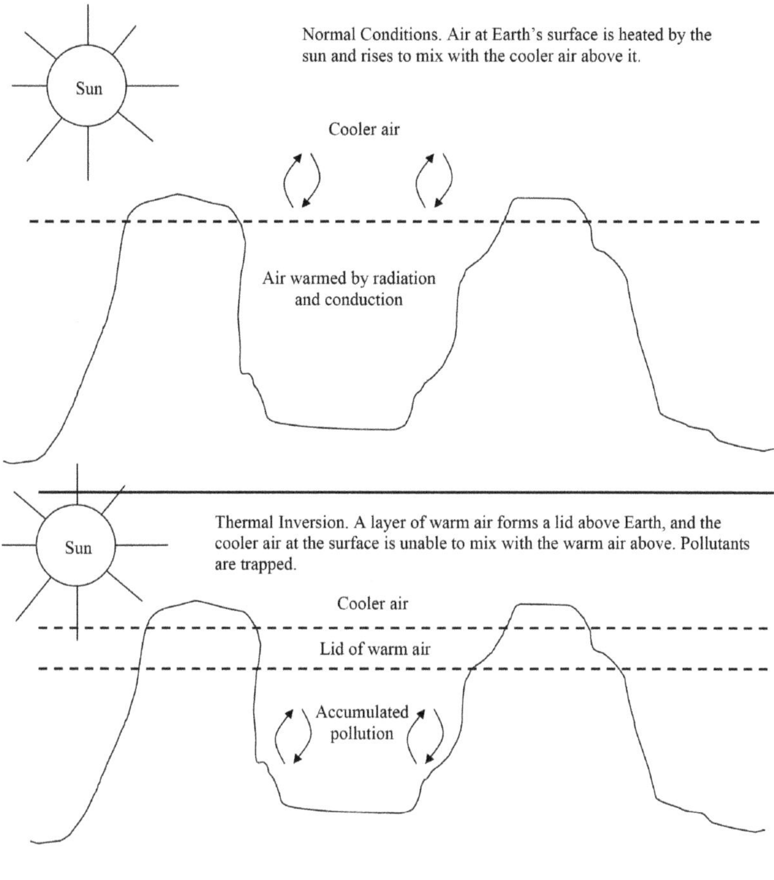

Figure 5.2. Thermal Inversion

automobiles—and Los Angeles possesses the ideal conditions for smog—conditions worsened by frequent thermal inversions.

WEATHER ELEMENTS[2]

What affects the weather? What causes rain, snow, hail storms, sunny days, cloudy days, windy days, and just plain old nasty

2. Adapted from NASA's How Atmospheric Pressure Effects the Weather. Accessed 06/27/11 @ http://nasa.gov/archive/air_pressure/barometer.html.

weather days? In answering these questions we begin by stating that weather is affected by certain weather elements. There are nine basic weather elements and they are defined as elements that can be measured in the atmosphere: air pressure, precipitation, wind, air temperature, relative humidity, solar energy, evaporation, pollution, and cloud height.

Air Pressure

Air pressure is the amount of force the atmosphere exerts on a surface. More specifically we can say that air pressure is the weight of the air over and around our bodies. It affects ensuing winds and the probability of rain. The most fundamental thing to know about air pressure is that heavier gases weigh more than lighter gases. So what? You ask. Well, different chemical elements, as you know, have different atomic weights. The chemical elements that form gases, like nitrogen, oxygen, etc. often combine two atoms at a time to form a gaseous molecule, like N_2 (two nitrogens) or O_2 (two oxygens).

The atomic weight of nitrogen (N) is 14 and of oxygen (O) is 16. The molecules N_2 and O_2 have molecular weights of 28 and 32, respectively. Obviously, a gallon of oxygen weighs more than a gallon of nitrogen. Moreover, it turns out that, at room temperature and normal (sea level) atmospheric pressure, 28 grams of nitrogen occupies a volume of 22.4 liters and 32 grams of oxygen occupies the same volume.

Recall that we stated that a weather element, such as air pressure, is something that we can measure in the atmosphere. Because air at a higher temperature is less dense than the cooler air over another air, it rises. This movement, in turn, generates difference in pressure between different areas. To measure the weather element pressure a barometer is used.

In regard to actual conditions, the pressure difference at the surface causes a wind to blow from the region of higher pressure toward that of lower pressure. It is similar to what happens when an inflated balloon is punctured at one end: The high-pressure air inside rushes out into the surrounding area of lower pressure. The greater the pressure difference, the stronger the wind. This principle operates on a global scale. Before hurricanes could be spotted by satellites, people would keep a wary eye on their barometers during hurricane season. If the air pressure dropped, that was usually a good time to board up windows and head further inland!

Warm air near the equator rises and moves toward the poles. In turn, the cooler polar air is drawn beneath, toward the equator. Keep in mind that air is always trying to equalize itself over the world, thus, as mentioned, is trying to leave an area of high pressure to get to a low-pressure area. The earth's rotation diverts these currents into prevailing easterly and westerly winds at different latitudes and altitudes.

These global movements of air masses in different directions generate turbulent motions where the main flow streams brush against each other. Topography also causes additional effects—the irregular outlines of continents and the complex patterns of mountains and plains, deserts, and forests.

When the weather changes, air pressure changes and causes the barometer to change. If the barometer is rising, the air is cooling and expanding and there is a tendency for fair weather. If it is dropping, the air is being compressed and is getting warmer so stormy weather could result.

Precipitation

One of the simplest weather elements to conceive (and measure) is precipitation. The official government definition is "all liquid or solid phase aqueous particles that originate in the atmosphere and fall to the earth's surface."

Wind

Wind is a weather element with two measurements that concern meteorologists: wind direction and wind speed. In regard to wind direction, we are all familiar with the weather vane, which is designed to turn itself into an oncoming wind; it tells us the direction the wind is coming *from*. Keep in mind that all wind direction information tells us where the wind is coming from, not where it is going.

Wind speed, or wind velocity, is a fundamental atmospheric rate and is measured by an anemometer; it can also be classified using the older Beaufort scale, which is based on people's observations of specifically defined wind effects.

Air Temperature

Temperature, another weather element, is a measure of how "hot" something is or how much thermal energy it contains. Specifically, it is the kinetic energy of the atmosphere and is measured with a thermometer. Temperature is a fundamental measurement in air science, especially in most pollution work. The temperature of a stack gas plume, for example, determines its buoyancy and how far the plume of effluent will rise before attaining the temperature of its surroundings. This, in turn, determines how much it will be diluted before traces of the pollutant reach ground level.

Temperature is measured on several scales: for example, the *centigrade* (*Celsius*) and *Fahrenheit* scales are both measured from a reference point—the freezing point of water—which is taken as 0°C or 32°F. The boiling point of water is taken as 100°C or 212°F. For thermodynamic devices, it is usual to work in terms of absolute or thermodynamic temperature where the reference point is absolute zero, which is the lowest possible temperature attainable. For absolute temperature measurement, the thermodynamic unit or *Kelvin* (K) scale is used, which uses centigrade divisions for which zero is the lowest attainable measurement. A unit of temperature on this scale is equal to a Celsius degree, but is not called a degree; it is called a Kelvin and is designated as K, not °K. The value of absolute zero on the Kelvin scale is –273.15°C, so that the Kelvin temperature is always a number 273 (rounded) higher that the Celsius temperature. Thus, water boils at 373 K and freezes at 273 K.

Relative Humidity

The amount of water in the air is called **humidity**. The weather element **relative humidity** is a dimensionless ratio, expressed in percent of the amount of atmospheric moisture (water vapor a true gas) present relative to the amount that would be present if the air were saturated. Because the latter amount is dependent on temperature, relative humidity is a function of both moisture content and temperature. As such, relative humidity by itself does not directly indicate the actual amount of atmospheric moisture present; moreover, it is difficult to measure accurately.

Warm air can hold more water than cold. When air with a given amount of water vapor cools, its relative humidity increases; when the air is warmed, its relative humidity decreases.

Solar Energy

The sun is the driving force behind weather. Without the distribution and reradiation to space of solar energy, we would experience no weather (as we know it) on Earth. The sun is the source of most of Earth's heat. Of the gigantic amount of solar energy generated by the sun, only a small portion bombards Earth—the rest is lost in space. A little over 40 percent of the sun's radiation reaching Earth hits the surface and is changed to heat. The rest stays in the atmosphere or is reflected back into space.

Like a greenhouse, the earth's atmosphere admits most of the solar radiation. When solar radiation is absorbed by the earth's surface, it is reradiated as heat waves, most of which is trapped by carbon dioxide and water vapor in the atmosphere, which work to keep the earth warm in the same way a greenhouse traps heat. (See the earlier chapter on greenhouse gases.)

The atmosphere plays an important role in regulating the earth's heating supply. The atmosphere protects the earth from too much solar radiation during the day and prevents most of the heat from escaping at night. Without the filtering and insulating properties of the atmosphere, the earth would experience severe temperatures similar to those of other planets.

On bright clear nights the earth cools more rapidly than on cloudy nights because cloud cover reflects a large amount of heat back to earth, where it is reabsorbed.

Evaporation

Evaporation is another weather element commonly used to measure weather. The NOAA (National Oceanic and Atmospheric Administration) explains that

> evaporation occurs when the physical state of water is changed from a liquid state to a gaseous state. A considerable amount of heat, about 600 calories of energy for each gram of water, is exchanged during the change of state. Typically, solar radiation and other factors such as air temperature, vapor pressure, wind, and atmospheric pressure affect the amount of natural evaporation that takes place in any geographic area.

Pollution

Air pollution affects weather directly and indirectly. For example, acid rain, global warming, and weather inversion pollution problems are all weather related, caused, at least to some degree, by air pollution.

Cloud Height

The ninth and final weather element is cloud height. **Cloud height** is the distance between the cloud base and the cloud top. Cloud height is often related to the intensity of precipitation generated by a cloud: deep clouds tend to produce more intense rainfall. For instance, cumulonimbus clouds can develop vertically through a substantial part of troposphere and often result in the thunderstorm with lightning and heavy showers.

WEATHER AS CLIMATE CHANGE INDICATOR

Now that the basic fundamentals of weather have been presented and explained it is time to shift our discussion to why they are major indicators of climate change. In light of this it is important to point out that rising global average temperature is associated with widespread changes in weather patterns. Scientific studies indicate that extreme weather events such as heat waves and large storms are likely to become more frequent or more intense with human-induced climate change. This section focuses on observed changes in temperature, precipitation, storms, floods, and droughts (USEPA, 2016).

The Temperature Indicator

This indicator of global climate change describes trends in average surface temperatures for the United States and the world. Warmer temperatures are one of the most direct signs that the climate is changing. Concentrations of heat-trapping greenhouse gases are increasing in the earth's atmosphere. In response, average temperatures at the earth's surface are increasing and are expected to continue rising. Note that because climate change can shift the wind patterns and ocean currents that drive the world's climate system, however, some areas are warming more than others, and some have experienced cooling (USEPA, 2016).

USEPA bases this indicator on daily temperature records from thousands of long-term weather monitoring stations, which have been compiled by the NOAA's National Centers for Environmental information. The indicator was developed by calculating annual anomalies, or differences, compared with the average temperature for 1901 to 2000. For example, an anomaly of + 2.0 degrees means the average temperature was 2 degrees higher than the long-term average. Each weather station has recorded daily and monthly anomalies. Global anomalies have been determined by dividing the world into a grid, averaging the data for each cell of the grid, and then averaging the grid cell together.

So, with regard to U.S. and global temperature trends let's look at the big picture. According to the NASA Goddard Institute for Space Studies, 2020 was the warmest year on record worldwide. "Overall, Earth's average temperature has risen more than 2 degrees Fahrenheit since the 1880s" (NASA, 2021).

The Precipitation Indicator

This indicator describes trends in average precipitation for the U.S. and the world, which can have wide-ranging effects on human well-being and ecosystems. Rainfall, snowfall, and the timing of snowmelt can all affect the amount of surface water and groundwater available for drinking, irrigation, and industry. They also influence river flooding and can determine what types of animals and plants (including crops) can survive in a particular place. A wide range of natural processes can be disrupted by changes in precipitation. This is particularly the case if these changes occur more quickly than plant and animal species can adapt. As average temperatures at the earth's surface rise more evaporation occurs, which in turn increases overall precipitation.

The result?

A warming climate is expected to increase precipitation in many areas. Just as precipitation patterns vary across the world, however, so will the precipitation effects of climate change. Some areas will experience decreased precipitation. Moreover, because higher temperatures lead to more evaporation, increased precipitation will not necessarily increase the amount of water available for drinking, irrigation, and industry.

On average, total annual precipitation has increased over land areas worldwide. Since 1901, global precipitation has increased at an average rate of 0.08 inches per decade (see figure 5.3). In the United States, on average total annual precipitation has increased over land areas. Some parts of the United States have experienced greater increases in precipitation than others. A few areas, such as the Southwest, have seen a decrease in precipitation. However, keep

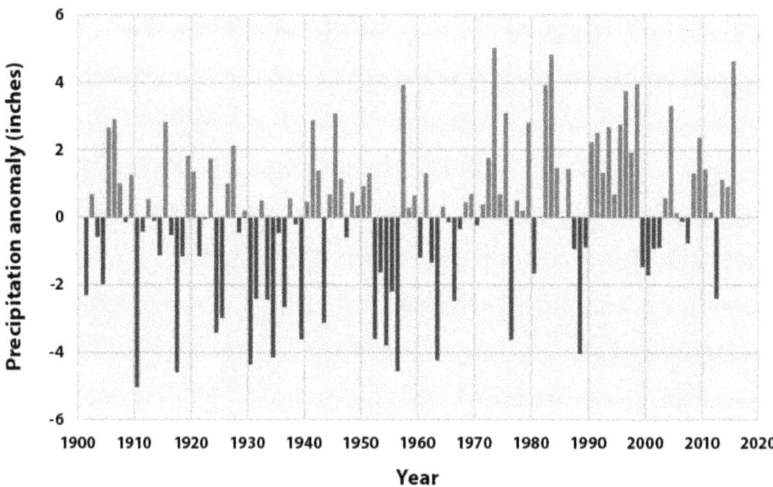

Figure 5.3. Change in Total Annual Amount of Precipitation over Land Worldwide since 1901
Source: USEPA (2016).

in mind that not all of these regional trends are statistically significant (USEPA, 2016).

Heavy Precipitation

This indicator tracks the frequency of heavy precipitation events in the United States. Heavy precipitation refers to instances during which the amount of rain or snow experienced in a location substantially exceeds what is normal. What comprises an episode of heavy precipitation varies according to location and season. Climate change can affect the intensity and frequency of precipitation. Warmer oceans increase the amount of water that evaporates into the air. When more moisture-laden air moves over land or converges into a storm system, it can produce more intense precipitation—for example, heavier rain and snowstorms (Melillo et al., 2014). The potential impacts of heavy precipitation include crop damage, soil erosion, and an increase in flood risk. In addition,

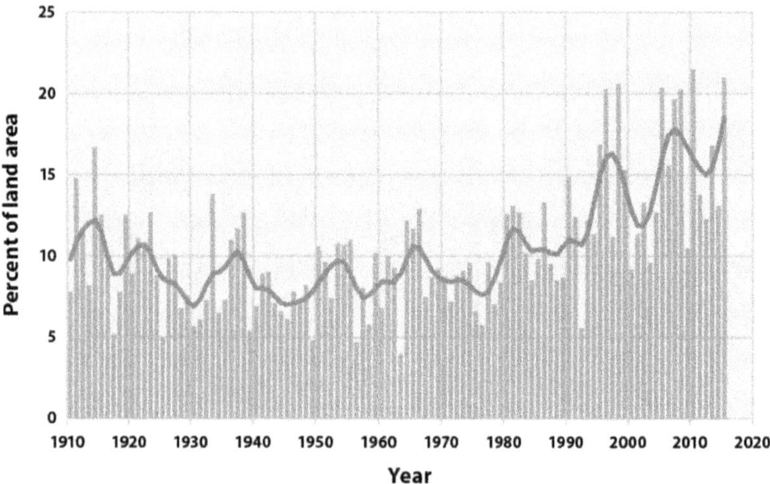

Figure 5.4. Land area of the contiguous 48 states where a much-greater-than-normal portion of total annual precipitation has come from extreme single-day precipitation events.

Note: The bars represent individual years, while the line is a nine-year weighted average.
Source: NOAA (2016).

runoff form precipitation can impair water quality as pollutants deposited on land wash into water bodies (see figure 5.4).

USEPA (2016) points out that in recent years, a larger percentage of precipitation has come in the form of intense single-day events. Nine of the top 10 years for extreme one-day precipitation events have occurred since 1990. The prevalence of extreme single-day precipitation events remained fairly steady between 1910 and the 1980s, but has risen substantially since then. Over the entire period from 1910 to 2015, the portion of the country experiencing extreme single-day precipitation events increased at a rate of about half a percentage point peer decade

Tropical Cyclone Activity

This indicator is about the frequency, intensity, and duration of hurricanes and other tropical storms in the Atlantic Ocean, Caribbean, and Gulf of Mexico. Hurricanes, tropical storms, and other intense rotating storms fall into a general category called cyclones. The historical record is replete with the known tropical cyclone events. The records show, for example, that sea storms disrupt and endanger shipping traffic. When cyclones encounter land, their intense rains and high winds can cause severe property damage, loss of life, soil erosion, and flooding. The associated storm surge—the large volume of ocean water pushed toward shore by the cyclone's strong winds—can cause severe flooding and destruction. Climate change is expected to affect tropical cyclones by increasing sea surface temperatures, a key factor that influences cyclone formation and behavior. The U.S. Global Change Research Program and the Intergovernmental Panel on Climate Change project that, more likely than not, tropical cyclones will become more intense over the twenty-first century, with higher wind speeds and heaver rains (Melillo et al., 2014; IPCC, 2013). Keep in mind that despite the apparent increases in tropical cyclone activity during recent years, changes in observation methods over time make it difficult to know whether tropical storm activity has actually shown a longer-term increase (IPCC, 2012).

River Flooding

This indicator examines changes in the size and frequency of inland river flood events in the United States. The natural result of large rainstorms or spring snowmelt that quickly drains into streams and rivers is flooding. It is possible that climate change causes these floods to become larger or more frequent than they used to be in terms of heavy precipitation events (Mallakpour and Villarini, 2015). Changes in streamflow, the timing of snowmelt, and the amount of snowpack that accumulates in the winter can also affect flood patterns. Although regular flooding helps to maintain the nutrient balance of soils in the flood plain, larger or more frequent floods could damage homes, roads, bridges, and other infrastructure; wipe out farmers' crops; harm or displace people; contaminate water supplies; and disrupt ecosystems by displacing aquatic life, impairing water quality, and increasing soil erosion.

Note that floods have generally become larger in rivers and streams across large parts of the Northeast and Midwest. Flood magnitude has generally decreased in the West, southern Appalachia, and northern Michigan.

Drought

This indicator measures drought conditions—a prolonged period of dry weather caused by a lack of precipitation that results from a serious water shortage for some activity, population, or ecological system—in the United States. Meteorologists also point out that drought can be thought of as an extended imbalance between precipitation and evaporation. Climate change has caused average temperatures to rise. With this temperature rise evaporation has increased, making more water available in the air for precipitation, but contributing to drying over some land areas and less moisture in the soil. Drought can have dramatic and disastrous effects. For example, drought conditions can negatively affect agriculture, water supplies, energy production and many other aspects of society. Lower streamflow and groundwater levels can also harm plants and animals, and dried-out vegetation increases the risk of wildfires (USEPA, 2016).

Looking at the historical record, note that average drought conditions across the nation have varied since records began in 1895. The 1930s and 1950s saw the most widespread droughts, while the last 50 years have generally been wetter than average. Looking at the period from 2000 through 2015, roughly 20 to 70 percent of the U.S. land area experienced conditions that were at least abnormally dry at any given time. Moreover, the years 2002–3 and 2012–13 had a relatively large area with at least abnormally dry contains, while 2001, 2005, and 2009–11 had substantially less area experiencing drought. Note that during the latter half of 2012, more than half of the U.S. land area was covered by moderate or greater drought. In several states, 2012 was among the driest years on record (NOAA, 2013).

DID YOU KNOW?

Rising temperatures and prolonged drought pose unique threats to indigenous populations because of their economic and cultural dependence on land and water supplies. Warming and drought can threaten medicinal and culturally important plants and animals, and can reduce water quality and availability, making tribal populations particularly vulnerable to waterborne illnesses (Gamble et al., 2016).

REFERENCES AND RECOMMENDED READING

Ackerman, D., 1990. *A Natural History of the Senses*. New York: Random House.

Bell, J. E., et al. 2016. Chapter 4: Impacts of Extreme Events on Human Health. The Impacts of Climate Change on Human Health in the United States: A Scientific Assessment, U.S. Global Change Research Program. https://health2016.global-change.gov.

Budyko, M. I. 1982. *The Earth's Climate*. New York: Academic Press.

Gamble, J. L., et al. 2016. Chapter 9: Populations of Concern. The Impacts of Climate Change on Human Health in the United States: A Scientific Assessment. U.S. Global Change Research Program. https://health2016.globalchange.gov.

Gates, D. M. 1962. *Energy Exchange in the Biosphere*. New York: Harper & Row Monographs.

Hesketh, H. E. 1991. *Air Pollution Control: Traditional and Hazardous Pollutants*. Lancaster, PA: Technomic Publishing Company.

IPCC (Intergovernmental Panel on Climate Change). 2012. Managing the Risks of Extreme Events and Disasters to Advance Climate Change Adaption. Cambridge, UK: Cambridge University Press.

IPCC. 2013. Climate Change 2013: The Physical Science Basis. Working Group I Contribution to the IPCC Fifth Assessment Report. Cambridge, UK: Cambridge University Press.

Mallakpour, I., and G. Villarini. 2015. The Changing Nature of Flooding across the Central United States. *Nature Climate Change* 5: 250–54.

Melillo, J. M, T. C. Richmond, and G. W. Yohe, eds. 2014. Climate Change Impacts in the United States: The Third National Climate Assessment, U.S. Global Change Research Program. http://nca2014.globalchange.gov.

Moran, J. M., and M. D. Morgan. 1994. *Essentials of Weather*. Upper Saddle River, NJ: Prentice-Hall.

NASA. 2007. *Pascal's Principle and Hydraulics*. Accessed December 29, 2018 @ www.grc.nasa.gov/WWW.

NASA. 2021. 2020 Tied for Warmest Year on Record, NASA Analysis Shows. January 14. https://www.nasa.gov/press-release/2020-tied-for-warmest-year-on-record-nasa-analysis-shows.

NOAA. 2007. *Weather Maps*. Accessed May 5, 2019 @ http://www.nssl.noaa.gov/edu/lessons/lesson_mpa.html.

NOAA. 2013. State of the Climate: Drought: December 2012. Accessed May 22, 2019, www.ncdc.noaa.gov/sotc/drought/2012/12.

NOAA. 2016. U.S. Climate Extremes Index. Accessed May 21, 2019. www.ncd.noaa.gov/extremes/cei.

National Research Council. 1982. *Solar Variability, Weather, and Climate*. Washington, DC: National Academy Press.

National Research Council. 1975. *Understanding Climatic Change, A Program for Action*. Washington, DC: National Academy of Sciences.

Slater, L., and G. Villarini. 2016. Update and Expansion to Data Originally Published in Mallakpour, I., G. Villarini, 2015. The Changing Nature of Flooding across the Central United States. *Nature Climate Change* 5: 250–54.

Spellman, F. R., and N. Whiting. 2006. *Environmental Science & Technology: Concepts and Applications*. Rockville, MD: Government Institutes.

USEPA. 2007. Basic Concepts in Environmental Sciences: Module 1 and 2. Accessed December 30, 2018 @ www.epa.gov/eogapti1/module 1 and 2.

USEPA. 2007. Air Pollution Control Orientation Course: Air Pollution. Accessed January 5, 2019 @ www.epa.gov/air/oaqps/eog/course422/ap.1.html.

USEPA. 2016. Climate change indicators in the United Sates, 2016. Fourth edition. EPA-430-R-16-004. www.epa.gov/climate-indicators.

Chapter 6

Climate Change Indicator Three

Oceans

> Atlantic Ocean circulation at weakest in a millennium, say scientists: Decline in system underpinning Gulf Stream could lead to more extreme weather in Europe and higher sea levels on US east coast.
>
> —*Guardian* (UK) headline, February 25, 2021

OCEANS SHAPE OUR CLIMATE ZONES

About 70 percent of Earth's surface is covered by oceans. Note that oceans have a two-way relationship with weather and climate. The oceans influence the weather on local to global scales; they shape our climate zones. At the same time, climate can fundamentally alter many properties of the oceans. In this chapter we examine how these important characteristics of oceans have changed over time.

At the present time greenhouse gases are trapping more energy from the sun; in turn, the oceans are absorbing more heat, resulting in an increase in sea surface temperatures and rising sea level. And it is important to point out that areas even hundreds of miles away from any coastline are still largely influenced by the global ocean (NOAA, 2019). There is no question about changes in ocean temperatures and currents (see figure 6.1) that are brought about by climate change and around the world. For example, warmer waters may

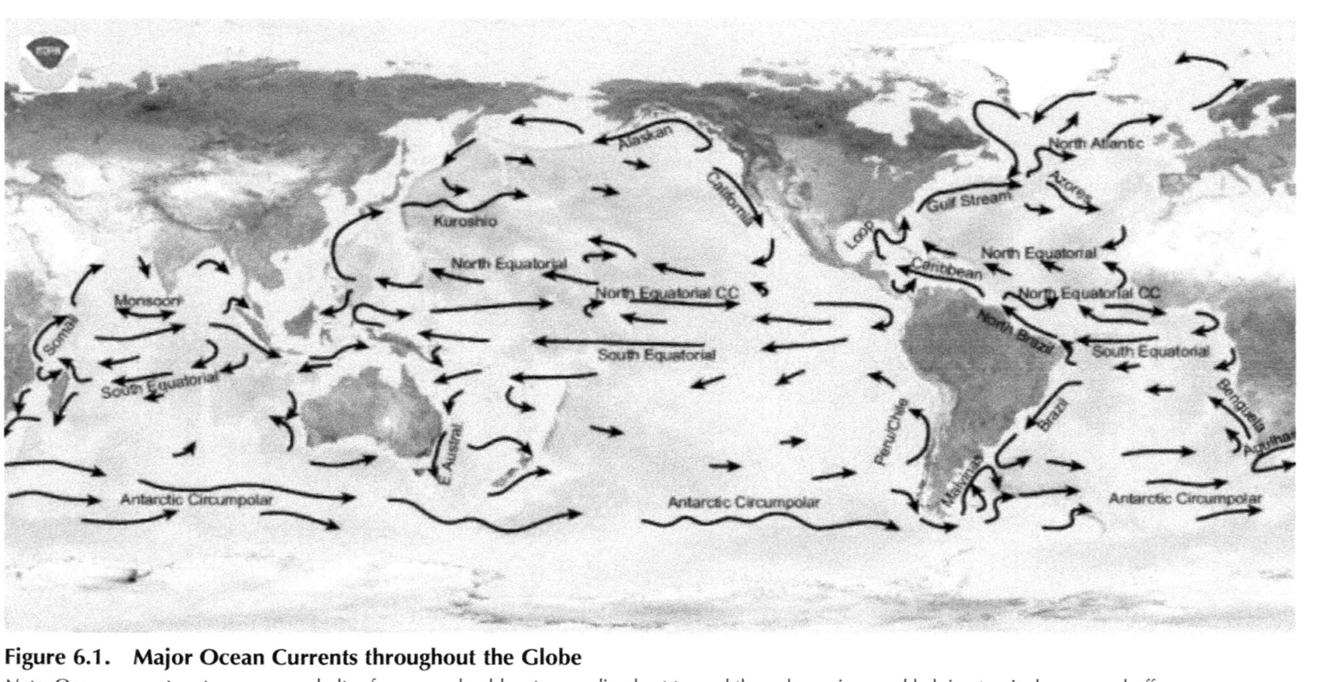

Figure 6.1. Major Ocean Currents throughout the Globe

Note: Ocean currents act as conveyor belts of warm and cold water, sending heat toward the polar regions and helping tropical areas cool off.

Source: NOAA (2019).

promote the development of stronger storms in the tropics, which can cause property damage and loss of life (USEPA, 2016). It is important to note that it is the coastal communities that are target one for the impacts associated with sea-level rise and stronger storms.

Although the oceans help mitigate climate change by storing large amounts of carbon dioxide, this process is a double-edged sword: one edge serves as a beneficial sink for storing carbon dioxide but the other edge is carbon dioxide's contribution to acidifying the oceans—changing the chemistry of seawater and making it more acidic. Increased ocean acidity makes it more difficult for certain organisms, such as shell fish and corals, to build their skeletons and shells. In turn, these effects could substantially alter the biodiversity and productivity of ocean ecosystems.

Changes in ocean systems do not occur overnight; instead, they occur over much longer time periods. This is the case because interactions between the oceans and atmosphere occur slowly over many months to years, and so does the movement of water within the oceans, including the mixing of deep and shallow waters. Trends can persist for decades, centuries, or longer. In contrast, in the atmosphere, storms can form quickly but can (and usually do) dissipate in a single day. The point is even if greenhouse gases were stabilized tomorrow, it would take many more years—decades to centuries—for the oceans to adjust to changes in the atmosphere and the climate that have already occurred (USEPA, 2016).

OCEAN HEAT

You might wonder where the human-caused warming ends up. Well, approximately 90 percent of the vast amount of human-caused warming ends up heating the oceans. Note that water has a tremendous capacity to store heat. Contrary to popular belief the atmosphere is a small repository to storing human-caused warming; it is the relatively poor heat storage capacity of the atmosphere that causes this phenomenon. Also, the earth's surface is not a large

DID YOU KNOW?

Ocean measurements determine the amount of heat energy stored in the ocean, based on measurements of ocean temperatures around the world at different depths. These measurements come from a variety of instruments deployed from airplanes and ships and, more recently, underwater robots. The point is the data collected are carefully adjusted to account for differences among measurement techniques and data collection programs. This indicator is based on analyses conducted by government agencies in three countries (USEPA, 2016).

storage warehouse of human-caused warming. The earth's surface simply can't store heat the way oceans do (Balmaseda et al., 2013). And because Earth's oceans store so much of the human-caused warming, this is a very important indicator (because we have seen rapid warming in the oceans).

Heating the Oceans

When sunlight reaches the earth's surface, the world's oceans absorb this energy as heat, which currents distribute around the world (see figure 6.1). As mentioned earlier, water has a much higher heat-storing capacity than air, meaning that oceans can absorb high amounts of heat with only a slight increase in temperature. Consequently, increasing concentrations of heat-trapping greenhouse gases have not caused the oceans to warm as much as the atmosphere, even though they have absorbed more than 90 percent of the earth's extra heat since 1955 (IPCC, 2013; Levitus et al., 2012). Note that if it were not for the large heat-storage capacity provided by the oceans, the atmosphere would grow warmer rapidly (Levitus et al., 2012). Water temperature reflects the amount of heat in the water at a particular time and location, and it plays an important role in the earth's climate system, because heat from ocean surface waters provides energy for storms, influences weather patterns, and changes ocean currents. Because water expands slightly as it gets

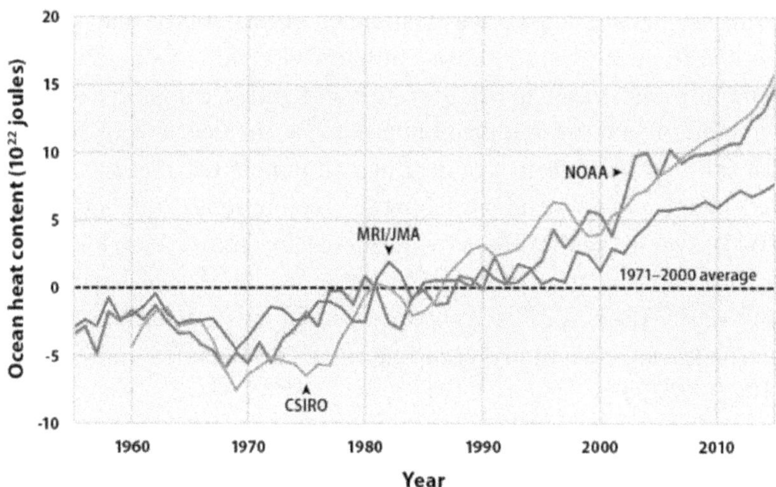

Figure 6.2. Changes in Ocean Heat Content, 1955–2015
Source: USEPA (2016).

warmer, an increase in ocean heat content will also increase the volume of water in the ocean, which is one cause of the observed increases in sea level.

In various studies and in three different data analyses, the long-term trend shows that the oceans have become warmer since 1955 (see figure 6.2).

Although concentrations of greenhouse gases have risen at a relatively steady rate over the past few decades, the rate of change in ocean heat content can vary from year to year. Year-to-year changes are influenced by events such as volcanic eruptions and recurring ocean-atmosphere patterns such as El Niño (USEPA, 2016).

SEA SURFACE TEMPERATURE

Global trends in sea surface temperature increase impact, and in many cases change ocean circulation patterns that transport warm and cold water around the globe, affecting which species are present in marine ecosystems, altering migration and breeding patterns, threatening corals (Ostrander et al., 2000), and changing the frequency and intensity of harmful algal blooms. Over the long term, increases in sea surface temperature could weaken the circulation patterns that bring nutrients for the deep sea to surface waters, contributing to declines in fish population that would affect people who depend on fishing for food or jobs (Pratchett et al., 2004). One of the problems with increasing sea surface temperature includes water vapor, which increases the risk of heavy rain and snow (IPCC, 2013). Another problem has to do with shifting storm tracks, potentially contributing to droughts in some areas (IPCC, 2013).

Looking at historical and current records as to what is happening to sea surface temperatures it is clear that surface temperature has increased during the twentieth century and continues to rise (see figure 6.3). From 1901 through 2015, temperature rose at an average rate of 0.13°F per decade. Sea surface temperature has been consistently higher during the past three decades than at any other time since reliable observations being in 1880. Again, based on historical record, increase in sea surface temperature has largely

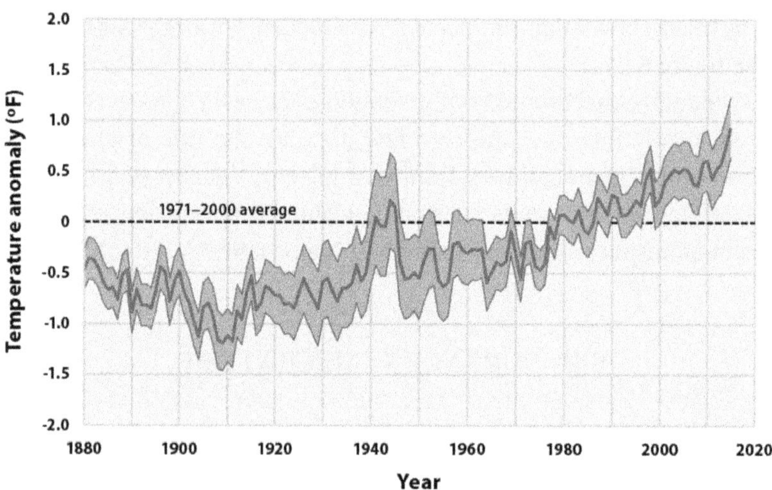

Figure 6.3. Change in Average Surface Temperatures of the World's Oceans since 1880

Source: NOAA (2016).

DID YOU KNOW?

A Climatological Study of the Effect of Sea-Surface Temperature on North Atlantic Hurricane Intensification found a statistically significant relationship between higher intensification values and higher sea surface temperature values.

occurred over two key periods: between 1910 and 1940, and from about 1970 to the present. Sea surface temperature appears to have cooled between 1880 and 1910.

SEA LEVEL

This indicator describes how sea level has changed over time; the change has come about due to changes in groundwater storage, glacier ice loss, Greenland ice loss, Antarctic ice loss, and the phenomenon of thermal expansion. Thermal expansion of water is the one phenomenon that has not changed over time—thermal expansion of water is a scientific fact, a normal, a reality, a truth; it raises sea levels because ocean water, like all water, expands as it warms up and thus takes up more space. Simply put, as the temperature of Earth changes, so does sea level; they are chained together, linked. Why? First, changes in the volume of water and ice on land (from snow banks, glaciers, and ice sheets) can increase or decrease the volume of water in the ocean. Second, as pointed out earlier, as water expands, it expands slightly—an effect that is cumulative over the entire depth of the oceans (USEPA, 2016). Low-lying wetlands and dry land are inundated by rising sea level. In addition, rising sea level erodes shorelines, contributes to coastal flooding, and increases the flow of salt water into freshwater estuaries and infiltrates nearby groundwater aquifers. Another huge vulnerability presents itself whenever high sea level combines with a strong storm to form a damaging surge that produces excessive and life-threatening flooding.

DID YOU KNOW?

Two different ways are used to measure sea level. *Relative* sea-level change refers to how the height of the ocean rises or falls relative to land at a particular location. In contrast, *absolute* sea-level change refers to the height of the ocean surface above the center of Earth, without regard to nearby land.

For approximately 2,000 years there was little change in global sea level. In the twentieth century, however, the rate of sea-level change accelerated (see figure 6.4; Titus et al., 2009). When averaged over all of the world's oceans, absolute sea level has risen at an average rate of 0.06 inches per year from 1880 to 2013. Since 1993, however, average sea level has risen at a rate of 0.11 to 0.14 inches per year—roughly twice as fast as the long-term trend.

Records reveal that relative sea level rose along much of the United States coastline between 1960 and 2015, particularly the

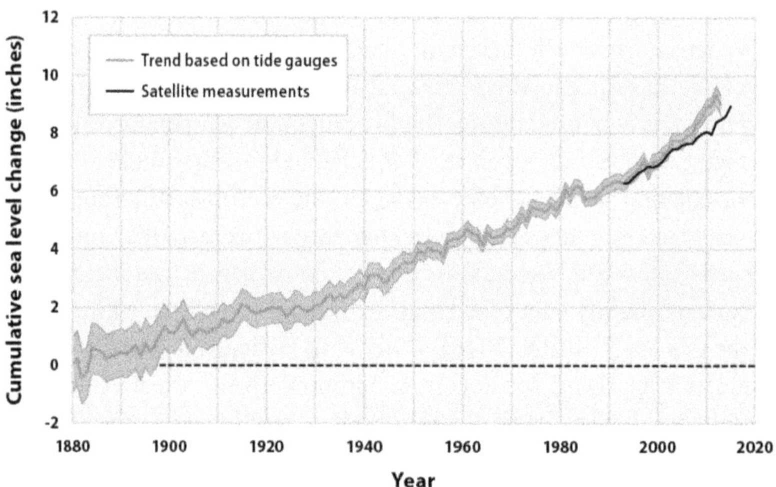

Figure 6.4. Cumulative Changes in Sea Level for the World's Oceans since 1880

Note: Data based on a combination of long-term tide-gauge measurement and recent satellite measurements.

Mid-Atlantic coast and parts of the Gulf Coast, where some stations registered increases of more than 8 inches. Meanwhile, relative sea level fell at some locations in Alaska and the Pacific Northwest. At those sites, even though absolute sea level has risen, land elevation has risen more rapidly. Note that relative sea level also has not risen uniformly because of regional and local changes in land movement and long-term changes in coastal circulation patterns (USEPA, 2016).

COASTAL FLOODING

The frequency of coastal flooding has changed over time. This occurrence is readily apparent to those who monitor tidal rise during high tide and during storm surges; moreover, it is most noticeable as sea level rises relative to the coast due to climate change, driving many coastal cities to define minor or "nuisance" flood thresholds. When water rises above this level, minor flooding typically occurs in some streets, many storm drains become ineffective (many were poorly engineered to handle the excess), and a coastal flood advisory may be issued, resulting in frequent road closures, reduced storm water drainage capacity, and deterioration of infrastructure not designed to withstand frequent inundation or exposure to salt water. Rising sea level increases the likelihood of flooding at high tide and during storm surges, and it also inundates low-lying wetlands and dry land, erodes shorelines, and increases the flow of salt water into estuaries and nearby groundwater aquifers. Millions of Americans and more than $1 trillion of property and infrastructure are at risk of damage from coastal flooding (Melillo et al., 2014).

Flooding is becoming more frequent along the U.S. coastline. As shown in figure 6.5, nearly every site measure has experienced an increase in coastal flooding since the 1950s. The rate is accelerating in many locations along the East and Gulf coasts. From figure 6.5 it is apparent that the Mid-Atlantic region suffers the highest number of coastal flood days and has also experienced the largest increases

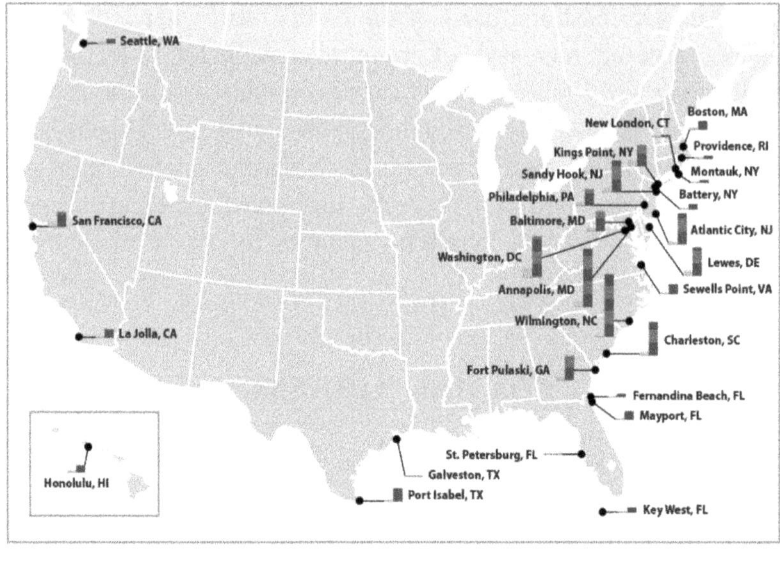

Figure 6.5. Average Number of Days per Year in Which Coastal Waters Rose above Local Thresholds of Minor Flooding at 27 Sites along U.S. Coasts
Source: NOAA (2016).

in flooding. Since 2010, Wilmington, North Carolina, has flooded most often—49 days per year—followed by Annapolis, Maryland, at 46 days per year. Annapolis, Wilmington, and two locations in New Jersey have also seen some of the most dramatic overall increases in frequency—floods are at least 10 times more common there than they were in the 1950s. The Mid-Atlantic's subsiding land and higher-than-average relative sea-level rise both contribute to this increase in flooding (USEPA, 2016).

OCEAN ACIDITY

Ocean acidity is an important indicator of climate change because the chemistry of the oceans is important in several ways. Ocean acidity is affected by the amount of carbon dioxide in the atmosphere. This is because as the concentration of carbon dioxide increase, the ocean absorbs more of it. During the past 250 years, oceans have absorbed about 28 percent of the carbon dioxide produced by human activities that burn fossil fuels (IPCC, 2013). Because dissolved carbon dioxide reacts with sea water to produce carbonic acid, marine life can be negatively affected. The increase in acidity changes the balance of minerals in the water and makes it more difficult for corals and plankton to produce the mineral calcium carbonate, which is the primary component of their hard skeletons and shells. Resulting declines in coral and plankton populations can change marine ecosystems and ultimately affect fish populations and the people who depend on them (Wooten et al., 2008). The problem is that signs of damage are already starting to appear in certain areas (Bednarsek et al., 2012).

DID YOU KNOW?

Carbon dioxide is measured as a partial pressure—a common way of measuring the amount of a gas. The actual acidity of ocean water is measured as pH—remember, lowering pH trends toward acidity.

REFERENCES AND RECOMMENDED READING

Balmaseda, M., et al. 2013. Distinctive Climate Signals in Reanalysis of Global Ocean Heat Content. *Geophys. Res. Lett.* 40, no. 9: 1754–59.

Bednarsek, N., G. A. Tarling, D. C. E. Bakkar et al. 2012. Extensive Dissolution of the Pheropods in the Southern Ocean. *Nat. Geosci.* 5: 881–85.

Durack, P., et al. 2014. Quantifying underestimates of long-term upper-ocean warming. *Nature Climate Change* 4: 999–1005.

IPCC. 2013. Climate Change 2013: The Physical Science Basis. Working Group I Contribution to the IPCC Fifth Assessment Report. Cambridge, UK: Cambridge University Press.

Levitus, S., et al. 2012. World Ocean Heat Content and Thermostatic Sea Level Change (0–2000 m), 1955–2010. *Geophys. Res. Lett.* 39: L10603.

Melillo, J. M, T. C. Richmond, and G. W. Yohe, eds. 2014. Climate Change Impacts in the United States: The Third National Climate Assessment, U.S. Global Change Research Program. http://nca2014.global change.gov.

NOAA. 2016. Extended Reconstructed Sea Surface Temperature. National Centers for Environmental Information. Accessed May 25, 2019 @ www.ncdc.noaa.gov/data-access/marine-ocean-data/extended-recon structed-sea-surface-temperature-ersst.

NOAA. 2019. Ocean Explorer: How Does the Ocean Affect Climate and Weather on Land? Accessed May 23, 2019 @ https://oceansexplorer .noaa.gov/facts/climate.html.

Ostrander, G. K., K. M. Armstrong, E. T. Knobbe, D. Gerace, and E. P. Scully. 2000. Rapid Transition in the Structure of a Coral Reef Community: The Effects of Coral Bleaching and Physical Disturbance. *P. Natl. Acad. Sci.* 97, no. 10: 5297–530.

Pratchett, M. S., S. K. Wilson, M. L Berumen, and M. J. McCormick. 2004. Sublethal Effects of Coral Bleaching on an Obligate Coral Feeding Butterfly Fish. *Coral Reefs* 23, no. 3: 352–56.

Titus, J. G., E. K. Anderson, D. R. Cahoon, S. Gill, R. E. Thieler, S. Williams. 2009. Coastal Sensitivity to Sea-Level Rise: A Focus on the Mid-Atlantic Region. U.S. Climate Change Science Program and the Subcommittee on Global Change Research. https://downloads.globalchnge .gov/sap/sap4-1-final-report-all.pdf.

USEPA. 2016. Climate Change Indicators in the United Sates, 2016. Fourth edition. EPA-430-R-16-004. www.epa.gov/climate-indicators.

Wootton, J. T., C. A. Plister, and J. D. Forester. 2008. Dynamic Patterns and Ecological Impacts of Declining Ocean pH in a High-Resolution Multi-year Dataset. *P. Natl. Acad. Sci.* 105, no. 48:18848–53.

Chapter 7

Climate Change Indicator Four

Snow and Ice

> Winter storm damage may rival Hurricane Harvey's price tag, experts say.
>
> —*Austin American-Statesman* headline, February 2021

SNOWMAGEDDON

During the month of February 2010, cold air in the wake of several reinforcing Arctic air masses dominated much of the United States, creating temperatures that were much below average in the Deep South. The combination of cold temperatures with the active weather pattern in the southern and eastern United States produced historic snowfall events. This historic event was dubbed "snowmaggedon" by many of the weather experts and news outlets. Simply, a *snowmaggedon* is a monumental winter storm lasting multiple days and resulting in copious amounts of snow and disastrous driving conditions. Thus, February 2010 was one of those events where excessive snow occurred, and the term was appropriately applied.

The record is clear: large snowstorm events have a huge impact on society in terms of economic loss, disruption, and human life. Examples include:

- Chicago blizzard of 1967 caused the deaths of 45 people and economic losses to local business estimated to be $150 million (Doesken and Judson, 1996).
- The 1993 "Superstorm" was responsible for 270 deaths and $1.8 billion in damages from the Deep South to New England (Kocin et al., 1995).
- In 1996, three large snowstorms struck the northeast causing $1.1 billion in insured losses (Kocin and Uccellini, 2005).
- In 2013, 10 snowstorms since 1980 caused damages totaling over $29 billion (Smith and Katz, 2013).
- Changnon (2007) reports that monetary losses resulting from snowstorms are increasing.
- Winter Storm Uri in February 2021 devasted the entire continental United States and parts of Canada and Mexico with unprecedented snowfall and widespread power outages. Damages are estimated at $19 billion. (Sechler and Hawkins, 2021)

These examples make it clear that we need to gain a better understanding of the impact of snowstorms.

DOMINANT FACTORS INFLUENCING U.S. WINTER STORMS

The two most dominant factors that influence United States winter characteristics (frequency, trajectory, intensity) are the El Niño–Southern Oscillation or ENSO and the North Atlantic Oscillation (NAO)/Arctic Oscillation (AO) phenomena.

El Niño–Southern Oscillation (ENSO)

El Niño–Southern Oscillation (ENSO) is a natural phenomenon that occurs every two to nine years on an irregular and unpredictable basis. El Niño is a warming of the surface waters in the tropical eastern Pacific that causes fish to disperse to cooler waters and, in turn, causes adult birds to fly off in search of new food sources elsewhere.

Through a complex web of events, El Niño (which means "the child" in Spanish and is named this because it usually occurs during the Christmas season off the coasts of Peru and Ecuador) can have a devastating impact on all forms of marine life.

During a normal year, equatorial trade winds pile up warm surface waters in the western Pacific. Thunderheads unleash heat and torrents of rain. This heightens the east-west temperature difference, sustaining the cycle. The jet stream blows from north Asia to California. During an El Niño–Southern Oscillation year, trade winds weaken, allowing warm waters to move east. This decreases the east-west temperature difference. The jet stream is pulled farther south than normal, picks up storms it would usually miss, and carries them to Canada or California. Warm waters eventually reach South America.

One of the first signs of its appearance is a shifting of winds along the equator in the Pacific Ocean. The normal easterly winds reverse direction and drag a large mass of warm water eastward toward the South American coastline. The large mass of warm water basically forms a barrier that prevents the upwelling of nutrient-rich cold water from the ocean bottom to the surface. As a result, the growth of microscopic algae that normally flourish in the nutrient-rich upwelling areas diminishes sharply, and that decrease has further repercussions. For example, El Niño–Southern Oscillation has been linked to patterns of subsequent droughts, floods, typhoons, and other costly weather extremes around the globe. Take a look at El Niño–Southern Oscillation's effect on the West Coast of the United States where ENSO has been blamed for West Coast hurricanes, floods, and early snowstorms. On the positive side, ENSO typically brings good news to those who live on the East Coast of the United States: a reduction in the number and severity of hurricanes.

Note that, in addition to reducing the number and severity of hurricanes, in October 1997 the Associated Press reported that ENSO also deserves credit for invigorating plants and helping to control the pollutants linked to global warming. Researchers have found that El Niño causes a burst of plant growth throughout the world and this removes carbon dioxide from the atmosphere.

Atmospheric carbon dioxide (CO_2) has been increasing steadily for decades. The culprits are increased use of fossil fuels and the clearing of tropical rainforests. However, during an ENSO phenomenon, global weather is warmer; there is an increase in new plant growth and CO_2 levels decrease.

Not only does ENSO have a major regional impact in the Pacific, its influence extends to other parts of the world through the interaction of pressure, air flow, and temperature effects.

El Niño–Southern Oscillation is a phenomenon that, although not quite yet completely understood by scientists, causes both positive and negative results, depending upon where you live.

North Atlantic Oscillation/Arctic Oscillation (NAO)/(AO)

The NAO/AO is a dominant influence on eastern United States weather patterns. The North Atlantic Oscillation—particularly in winter—is a major mode of atmospheric variability in the Northern Hemisphere. The different phases of the NAO result in different winter conditions over the north Atlantic, with a positive NAO index rendering more and stronger wind storms crossing the Atlantic Ocean on a more northerly track; however, with a negative NAO index, fewer and weaker winter storms cross on a more west-east pathway.

The Arctic Oscillation (AO) index describes the relative intensity of a semipermanent low-pressure center over the North Pole. A vortex is formed by a band of upper-level winds circulating around the center. When the vortex is strong and tight it traps cold air around the North Pole. However, when the vortex is weak it allows cold air intrusions plunging southward into North America, Europe, and Asia.

SNOW AND ICE

As the climate has changed, years with heavy seasonal snow and extreme snowstorms continue to occur with great frequency. The frequency of extreme snowstorms in the eastern two-thirds of the contiguous United States has increased over the past century. Approximately twice as many extreme United States snowstorms occurred in the latter half of the twentieth century than the first (NOAA, 2019a).

The earth's surface contains many forms of snow and ice, including sea, lake, and river ice; snow cover; glaciers, ice caps, and ice sheets; and frozen ground. Climate change can dramatically alter the earth's snow- and ice-covered areas because snow and ice can easily change between solid and liquid states in response to relatively minor changes in temperature.

Despite the occurrence of extreme snowstorms, there is still a danger of overall reduction in snow cover. What does this mean? Reduced snowfall and less snow cover on the ground could diminish the beneficial insulating effects of snow for vegetation and wildlife, while also affecting water supplies, transportation, cultural practices, travel, and recreation for millions of people (USEPA, 2016). For communities in Arctic regions, reduced snow and ice could present commercial opportunities for those using shipping lanes that are ice free. Low levels of snow and ice could also benefit those seeking access to natural resources. Conversely, reduced sea ice could increase coastal erosion and exposure to storms, threatening property, while thawing ground could damage roads and buildings and accelerate erosion (USEPA, 2016).

REGIONAL SNOWFALL INDEX (RSI)

In the United States, how are snowstorm events ranked? Where are the regional areas in the United States where snowstorm events are ranked? Who does the ranking? What kind of ranking scale is used?

Table 7.1. Regional Snowfall Index (RSI)

Category	*RSI Value*	*Description*
1	1–3	Notable
2	3–6	Significant
3	6–10	Major
4	10–18	Crippling
5	18.0+	Extreme

Source: NOAA (2019b).

These are all good questions. In the United States, snowstorm events are ranked to the Regional Snowfall Index (RSI). The areas ranked occur in and impact the eastern two thirds of the United States. The ranking of snowstorm events is produced by NOAA's National Centers for Environmental Information. The RSI ranks snowstorm impact on a scale from 1 to 5, similar to the Fujita scale for tornadoes or the Saffir-Simpson scale for hurricanes. The RSI differs from these other indices because it includes population. RSI is based on the spatial extent of the storm, the amount of snowfall, and the juxtaposition of these elements with population, including population information tying the index to societal impacts. Currently, the index uses population based on the 2000 Census (NOAA, 2019b). The RSI ranking scale is shown in table 7.1.

The bottom line: Changing climate conditions can have worldwide implications because snow and ice influence air temperatures, sea level, ocean currents, and storm patterns. For example, melting ice sheets on Greenland and Antarctica add fresh water to the ocean, increasing sea level and possibly changing ocean circulation driven by differences in temperature and salinity. Because of their light color, snow and ice also reflect more sunlight than open water or bare ground, so a reduction in snow cover and ice causes the earth's surface to absorb more energy from the sun and become warmer (USEPA, 2016).

ARCTIC SEA ICE

When we think about the Arctic we often think about or visualize sea ice. Because of global climate change scientists (and others) have been paying increased attention to the Arctic Region. Experts are currently tracking the extent, age, and melt season of sea ice in the Arctic Ocean. In tracking Arctic sea ice, attention is being paid to annual occurrences. For example, each year some of the ice melts during the summer because of warmer temperatures and sunlight, typically reaching its minimum thickness and extent in mid-September. In the fall, the ice freezes and begins expanding again.

Sea ice extent is an important indicator of global climate change because warmer air and water temperatures are reducing the amount of sea ice present. Sea ice reflects sunlight, which helps to keep polar regions cool. Sea ice is also important because it provides habitat for animals such as the tiny shrew (as small as a bumblebee), 800-pound moose, birds of all descriptions, freshwater and marine fish, numerous insects, walruses, polar bears, caribou, and reindeer. Caribou and wild reindeer populations have dropped 56 percent in a 20-year period (NOAA, 2018). Also, wildlife and ice travel are vital to the traditional subsistence lifestyle of indigenous Arctic communities.

THE ARCTIC REPORT CARD

So, what's really happening to and in the Arctic? Well, looking at the record, note that September 2012 had the lowest sea ice extent ever recorded, for that month 44 percent below the 1981–2010 average. Later, in 2015, sea ice extent was more than 700,000 square miles less than the historical 1981–2010 average for that month—a difference more than two and a half times the size of Texas. March sea ice reached the lowest extent on record in 2015 and hit roughly the same low again in 2016—about 7 percent less than the 1981–2010 average (USEPA, 2016).

GREENLAND: THE CANARY IN A COAL MINE

In centuries past, coal miners, who had no ventilation air supply, used caged canaries lowered into shafts to detect any dangerous gas build-up—methane or carbon monoxide. Today, with regard to climate change the "canary" used by climate watchers is Greenland. More specifically, it is the Greenland Ice Sheet (NOAA, 2018) that is the canary in the coal mine. (See figures 7.1–7.3.)

When we look at the record, in a stitch in time (2018), of surface melting, surface ice mass balance, total ice mass balance, albedo, surface air temperature, and marine terminating glaciers of the Greenland ice sheet we find a mixed bag. The 2018 record, the most recent record, shows that estimates of the spatial extent of melt across the Green ice sheet were unexceptional for most of the summer melt season (June, July, and August) of 2018. In addition, surface ice mass balance for the 2017/2018 season was below of near the long-term mean (relative to the period 1961–90) consistent with snow cover that survived late into the spring and average/low surface melting during summer. Summer 2018 albedo (a measure of surface reflectivity) average over the whole ice sheet, tied with the record high set in 2000 for the 2000–18 period. Relatively high albedo was associated with the reduced surface melting and extended survival of the snow cover, which reduced the exposure of darker, bare ice.

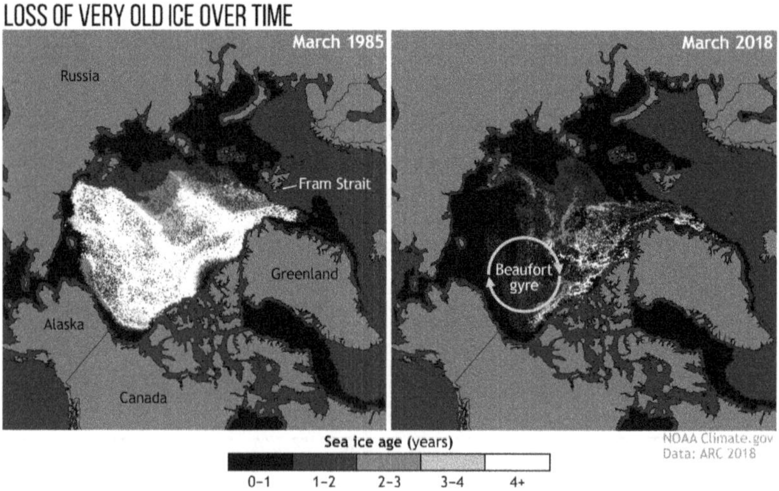

LOSS OF VERY OLD ICE OVER TIME

March 1985

Russia

Fram Strait

Greenland

Alaska

Canada

March 2018

Beaufort gyre

Sea ice age (years)

0–1 1–2 2–3 3–4 4+

NOAA Climate.gov
Data: ARC 2018

Figure 7.1. Declining Sea Ice

Note: The 2017 Arctic Report Card found the Arctic region had the second-lowest overall sea-ice coverage on record.

Source: NOAA (2018).

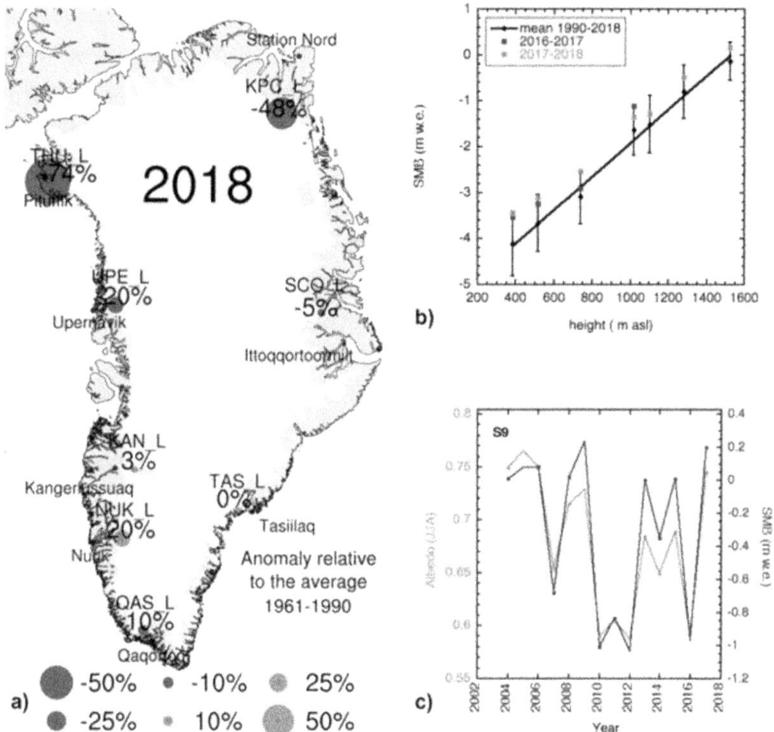

Figure 7.2. Greenland Ice Sheet Ablation Anomalies, 2017
Source: NOAA (2018).

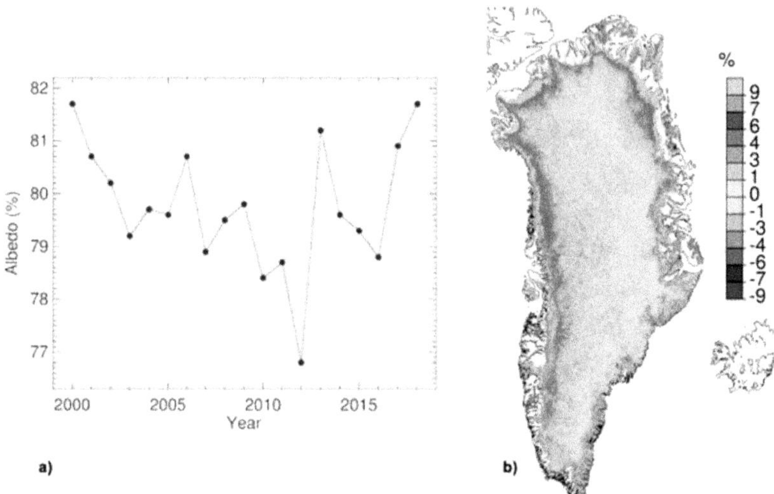

Figure 7.3. Time Series Summer Albedo Average over the Entire Greenland Ice Sheet, 2017
Source: NOAA (2018).

It can't be overstated about the importance of the Greenland ice sheet and the role it plays in gauging the health of our planet. The high albedo of the ice sheet contributes to modulating the amount of new solar energy absorbed by the earth and controls atmospheric circulation because of its location and topography. On the flip side, the data show that surface air temperatures set new high records in winter (up to +14.4°C above the mean) and low records in summer (−46.3°C) (NOAA, 2018).

Evidence of the age of Arctic sea ice suggests that fewer patches of ice are persisting for multiple years (i.e., generally thick ice that has survived one or more melt seasons). The proportion of sea ice five years or older has declined dramatically over the recorded time period, from more than 30 percent of September ice in the 1980s to 9 percent in 2015. A growing percentage of sea ice is only one or two years old. Less old multiyear ice implies that the ice cover is thinning, which makes it more vulnerable to further melting. Since 1979, the length of the melt season for Arctic sea ice has grown by 37 days. Arctic sea ice now starts melting 11 days earlier and it starts refreezing 26 days later than it used to, on average (USEPA, 2016).

ANTARCTIC SEA ICE

Like Arctic sea ice, Antarctic sea ice affects global climate, regional climate, and ecosystems. But Antarctica and the Arctic are two very different environments: the former is a continent surrounded by ocean, enclosed by land. Consequently, sea ice behaves differently in the two regions.

The Southern Ocean around Antarctica freezes to form sea ice every year. This sea ice reaches its maximum extent in September or early October and melts in the summer months (December to February). The sea ice around Antarctica is thinner, and nearly all of it melts in a typical summer. Antarctic sea ice is influenced by wind patterns, ocean currents, and precipitation around the continent (IPCC, 2013).

The Antarctic sea ice yearly wintertime maximum hit record highs from 2012 to 2014 before returning to average levels in 2015. On September 19, 2014, the five-day average of Antarctic sea ices extent exceeded 20 million square kilometers for the first time since 1979, according to the National Snow and Ice Data Center. In contrast, the Arctic wintertime maximum and its summer minimum have been in sharp decline for the past decade. Studies show that globally, the decreases in Arctic sea ice far exceed the increases in Antarctic sea ice (NASA, 2017).

GLACIERS

Glaciers are large masses of snow and ice that have accumulated over thousands of years and are present year-round in many areas. Glaciers not only transport material as they move, but they are also sculptors, carving away the land beneath them. A glacier's gradual movement, combined with its weight and its sharp and skilled "hands," so to speak, can drastically reshape the landscape over hundreds or even thousands of years. The ice sculpts the land surface and carries the broken rocks and soil debris far from their original locations, resulting in some spectacular glacial landforms, like the Great Lakes, for example.

Along with the Great Lakes, probably the most readily visible glacial land forms are glaciated valleys, which are common all over the world. Glaciers also provide localities and ecosystems with a reliable source of streamflow and drinking water particularly in the late summer and in times of drought, when seasonal snowpacks have melted away.

Glaciers—wrapped in majesty—fascinate us. When we look upon, photograph, or hike among existing glaciers, we're in awe. But they can tell us so much. Consider one example: the Matanuska Glacier, the largest and most spectacular sight between the Alaskan cities of Palmer and Glennallen (see figures 7.4 and 7.5). It is currently 27 miles long and 4 miles wide.

Figure 7.4. Matanuska Glacier, Alaska, 2010
Source: F. R. Spellman and JoAnn Chapman.

Figure 7.5. Matanuska Glacier, Alaska, 2010
Note: Compare the immensity of the glacier to the climber at top.
Source: F. R. Spellman and JoAnn Chapman.

Glaciers are important indicators of climate change. Their growth or shrinkage, their advance or retreat—these are the factors that provide evidence of changes in precipitation and temperature. If glaciers lose more ice than they can accumulate through new snowfall, they ultimately add more water to the oceans, leading to a rise in sea level.

Glaciers have changed over time. Presently, many glaciers are shrinking or have disappeared completely.

Since at least the 1970s, glaciers worldwide (on average) have been losing mass, which in turn has contributed to observed changes in sea level. A longer measurement record from a smaller number of glaciers suggests that they have been shrinking since the 1940s. Note that the rate at which glaciers are losing surface mass balance appears to have accelerated over roughly the last decade (USEPA, 2016).

Okay, glaciers are losing mass balance, what does that mean?

First, let's talk about the USGS (United States Geological Survey) and other glacier experts using four benchmark glaciers in the United States—the Wolverine Glacier, South Cascade Glacier, Gulkana Glacier, and Sperry Glacier. These glaciers have shown an overall decline in mass balance since the 1950s and 1960s and an accelerated rate of decline in recent years.

Wolverine Glacier

The Wolverine Glacier in southeastern Alaska (see figure 7.6) has been measured by the USGS since 1966; it is located in a maritime climate in the Kenai Mountains. The Wolverine Glacier is one of two benchmark glaciers in Alaska where its surface mass balance has been closely monitored. Repeated measurements at three long-term "index" sites, in conjunction with local meteorology and runoff data, are used to estimate glacier-wide mass balances. These data constitute the longest continuous set of mass-balance data in North America (Josberger et al., 2007), which are used to understand glacier dynamics and hydrology and to understand the glacier's response to climate change (USGS, 2019a).

Figure 7.6. Wolverine Glacier, Alaska
Source: USGS (2019a).

South Cascade Glacier

South Cascade Glacier is located in a north-northwest facing valley near the crest of the Cascade Range in northwest Washington State. The glacier receives heavy annual precipitation, exceeding 4.5 m in places (USGS, 2019b) and is heavily influenced by climate conditions in the Pacific Ocean—the mass balance of glaciers depend on precipitation and temperature in winter and on temperature and insolation (i.e., exposure to the sun's rays) in summer (Walters and Meier, 1989). In 2007, South Cascade glacier had an area of 1.73 km^2. The glacier spans nearly 660 m in elevation in the steep side basin, which forms the headwaters of the South Fork of the Cascade River.

In 1959 scientists with the USGS began collecting mass balance data on South Cascade Glacier. Data collected are based on measurements including streamflow runoff, precipitation, air temperature, barometric pressure, snow thickness and density, ice ablation, surface speed, and surface altitude (USGS, 2019b).

Gulkana Glacier

Gulkana Glacier is located in the central Alaska Range in interior Alaska. The climate is continental, characterized by a large range in temperatures and irregular, lighter precipitation as compared to more coastal areas. The main trunk of the glacier has a southerly aspect, and is fed by four cirques (i.e., amphitheater-shaped hollows) with west, south, and east aspects.

In 1966 scientists with the USGS began collecting mass balance data on Gulkana Glacier, the second of the two benchmark glaciers in Alaska. Repeated point measurements at established sites, in conjunction with local meteorology and runoff data, are used to estimate glacier-wide mass balances. These data constitute the longest continuous set of mass-balance data in North America (Josberger et al., 2007), which are used to understand glacier dynamic and hydrology and to understand the glacier's response to climate change (USGS, 2019c).

Sperry Glacier

The final of the four benchmark glaciers, Sperry Glacier, occupies a broad, shallow cirque situated just beneath and west of the continental Divide in the Lewis Mountain Range of Glacier National Park, Montana (see figure 7.7). Sperry Glacier is wider than it is long relative to its flow direction and spans about 300 m in elevation with a median altitude of 2450 m.

As a result of its position on the Continental Divide the glacier is influenced by both maritime and continental air masses. However, given its position on the western and predominantly windward side of the Continental Divide, Pacific storm systems dominate the weather. These bring heavy precipitation and moderate temperatures as warm, moist Pacific fronts collide with and lift over the Rocky Mountains. Temperature and precipitation patterns in northwest Montana are marked by strong altitudinal gradients. For valley sites

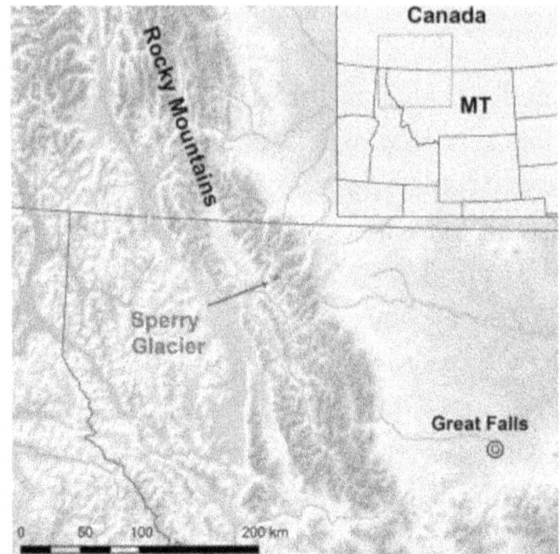

Figure 7.7. Sperry Glacier, Montana
Source: USGS (2019c).

at about 1000 m, mean temperatures for July, which is generally the warmest month of the year, are typically 58–63°F (15–17°C); they are roughly half that for mountain sites at 2500 m (USGS, 2019c).

Sperry became the focus of extensive field research starting in 2005 as scientists employed standard glaciology methods (Ostrem and Brugman, 1991) to estimate glacier-wide seasonal and annual surface mass balances. Snow depths and densities are measured in the spring when the glacier's balance is at a maximum. Ablation stakes are also installed at this time and then checked periodically during the summer melt season with a final check in early autumn at the balance minima. The Sperry Glacier joined the established USGS Benchmark Program in 2013 (USEPA, 2016).

DID YOU KNOW?

Glacial ablation is the loss of ice and snow from a glacier system. This occurs through a variety of processes including melting and runoff, sublimation, evaporation, calving, and wind transportation of snow off of a glacier basin.

Monitoring and measuring benchmark glaciers allows scientists to interpret the dynamics involved with surface mass balance manifestations based on climatic activities. Simply, if cumulative mass balance becomes more negative over time, it means glaciers are losing mass more quickly than they can accumulate more snow and ice, which can help to determine climate change effects.

LAKE ICE

Lake ice formation, thaw, and breakup dates—and when open water becomes extensive—are key indicators of climate change (USEPA, 2016). This indicator measures the amount of time that ice is present on lakes in the United States. If lakes remain frozen for longer periods, it can signify that the climate is cooling. Conversely, shorter periods of ice cover suggest a warming climate. Note that changes in ice cover can affect the physical, chemical, and biological characteristic of a body of water. Reduced ice cover leads to increased evaporation and lower water levels, as well as an increase in water temperature and sunlight penetration, which in turn can affect plants and animals. The timing and duration of ice cover on lakes and other bodies of water can also affect society—particularly in relation to shipping and transportation, hydroelectric power generation, and fishing.

Research via monitoring and measuring has shown that lakes covered by this indicator are generally freezing later than they did in the past. Freeze dates have shifted later at a rate of roughly half a day to one-and-a-half days per decade. Moreover, thaw dates for most of these lakes show a trend toward earlier ice breakup in the spring. Spring thaw dates have gown earlier by up to 24 days in the past 110 years (see figure 7.8). All of these lakes were found to be thawing earlier in the year. The changes in lake freeze and thaw dates shown in figure 7.8 are consistent with other studies. Note that when a lake freezes, snowfall piles up with the occurrence of winter snowstorms.

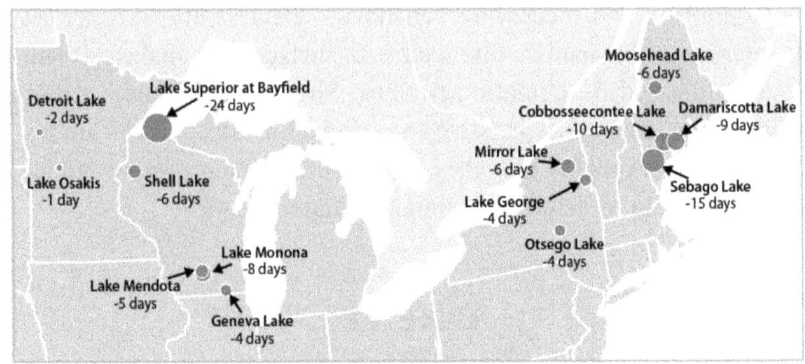

Change in ice thaw date:

Earlier

Figure 7.8. "Ice Out" Dates (or Dates of Ice Thawing and Breakup) for 14 U.S. Lakes, 1905–2015

Note: Large circles indicate larger changes; other circles represent negative numbers with earlier thaw dates.

Source: Hodgkins (2010); USGS (2016); Minnesota Department of Natural Resources (2019); Geneva Lake Environmental Agency (2016); North Temperate Lakes Long-Term Ecological Research (2016); Adirondack Daily Enterprise (2013); Lake Placid News (2016); State University of New York (2015); National Snow and Ice Data Center (2014).

There's an annual competition between the communities of Nenana on the Tanana River in Alaska and the Yukon River in Dawson City, Yukon, to guess (with wagering) ice break day and time. A reward goes to the wining guesser. This annual guessing and wagering have been taking place for more than 100 years. Detailed records from the competition have shown that both the Tanana and Yukon rivers demonstrate long-term trends toward earlier ice breakup in the spring. The ice breakup dates for both the Tanana and Yukon rivers have shifted earlier by approximately seven days over their respective periods of record. In addition, in 2016 the breakup date recorded at Dawson City was the earliest on record. Still, it is important to point out that despite the overall trend toward earlier breakup, recent breakup dates for both rivers are largely within the range of historical variation (USEPA, 2016; Melillo, Richmond, and Yohe, 2014; Beltaos and Burrell, 2003; Nenana Ice Classic, 2016; Yukon River Breakup, 2016).

SNOWFALL

This indicator uses two different measures to show how snowfall has changed in the contiguous 48 states. Snowfall is important to many of us; it is an important aspect of winter in much of the United States. Warmer temperatures cause more water to evaporate from the land and oceans, which leads to more precipitation, larger storms, and more variation in precipitation in some areas. Generally speaking a warmer climate will cause more of this precipitation to fall in the form of rain instead of snow. Changes in the amount and timing of snowfall could affect the spawning of fish in the spring and the amount of water available for people to use in the spring and summer. With winter recreational activities, a lack of enough snowfall can have an impact on such businesses as ski resorts and outdoor fun.

DID YOU KNOW?

Total snowfall has decreased in many parts of the country since widespread observations became available in 1930, with 57 percent of stations showing a decline. Among all of the stations shown, the average change is a decrease of 0.19 percent per year (USEPA, 2016).

It is important to point out that in addition to changing the overall rate of precipitation, climate change can lead to changes in the *type* of precipitation. One reason for the decline in total snowfall is because more winter precipitation is falling in the form of rain instead of snow. Nearly 80 percent of the stations across the contiguous 48 states have experienced a decrease in the proportion of precipitation falling as snow (USEPA, 2016).

Varying snowfall trends are regional. The Pacific Northwest has seen a decline in both total snowfall and the proportion of precipitation falling as snow. Parts of the Midwest have also experienced a decrease, particularly in terms of the snow-to-precipitation ratio.

On the other hand, a few regions have seen modest increases, including some areas near the Great Lakes that now receive more snow than in the past.

SNOW COVER

So, what is the amount of land in North America that is covered by snow? Good question. To answer this question climatologists and others use the term **snow cover**. Simply, the parameter snow cover refers to the amount of land covered by snow at any given time, which is influenced by the amount of precipitation that falls as snow. As temperature and precipitation patterns change, so can the overall area covered by snow. Snow cover is not just something that is affected by climate change, however; it also *exerts* an influence on climate. Recall that in table 4.2 the percentage of albedo (reflectivity of Earth's surface) for various surface types is presented and lists that fresh snow has an albedo of 75 to 95 percent, old snow is at 40 to 70 percent, glacier ice is 20 to 40 percent, and sea ice is 30 to 40 percent. The point is that more snow cover means that more energy reflects back to space, resulting in cooling, while less snow cover means more energy is absorbed at the earth's surface, resulting in warming.

When averaged over the entire year, snow covered an average of 3.24 million square miles of North America during the period from 1972 to 2015. The fact is the extent of snow cover has varied from year to year. The average area covered by snow has ranged from 3.0 million to 3.6 million square miles, with the minimum value occurring in 1998 and the maximum in 1978. Between 1972 and 2015, the average extent of North American snow cover decreased at a rate of about 3,300 square miles per year. The average area covered by snow during the most recent decade (2006–15) was 3.21 million square miles, which is about 4 percent smaller than the average extent during the first 10 years of measurement (1972–81)—a difference of 122,000 square miles, or approximately an area the size of New Mexico (USEPA, 2016).

Note that since 1972, the U.S. snow cover season has become shorter by nearly two weeks, on average. By far the largest change has taken place in the spring, with the last day of snow shifting earlier by 19 days since 1972. In contrast, the first date of snow cover in the fall has remained relatively unchanged.

SNOWPACKS

Another significant indicator of climate change is snow accumulation; that is, **snowpacks** (accumulations) consisting of layers of snow in certain geographic locations and high latitudes where the climate includes cold weather for extended periods during the year, measuring trends in mountain snow accumulation in the western United States. Snowpacks are an important water resource that feed streams and rivers as they melt. As a result, snowpacks are both the drinking water source for many communities and a potential source of flooding (in case of sudden melting). Snowpacks also contribute mass to glaciers in their accumulation zone.

Keep in mind that it is temperature and precipitation that are the key factors affecting snowpack. Mountain snowpack plays a key role in the water cycle in western North America. Millions of people in the West depend on the melting of mountain snowpack for power, irrigation, and as stated, drinking water. Changes in mountain snowpack can affect agriculture, winter recreation, and tourism in some areas, as well as plants and wildlife. In a warming climate that is obvious and apparent at present, more precipitation is expected to fall as rain rather than snow in most areas—reducing the depth and extent of snowpack. Plainly, higher temperatures in the spring can cause snow to melt earlier.

Large and consistent decreases have been observed throughout the western United States. Decreases have been especially prominent in

Washington, Oregon, and the northern Rockies. Consider that from 1955 to 2016, April snowpack declined at more than 90 percent of the sites measured. The average change across all sites amounts to about a 23 percent decline. While it is true that some stations have seen increases in snowpack, all 11 states included in this indicator have experienced a decrease in snowpack on average over the time period. In the Northwest (Idaho, Oregon, Washington), all but three stations saw decreases in snowpack over the period of record (USEPA, 2016).

REFERENCES AND RECOMMENDED READING

Adirondack Daily Enterprise. 2013. https://www.adirondackdailyenterprise.com.

Beltaos, S., and B. C. Burrell. 2003. Climatic Change and River Ice Breakup. *Can. J. Civil Eng.* 30: 145–55.

Changnon, S. A. 2007. Catastrophic Winter Storms: An Escalating Problem. *Climatic Change* 84: 131–39. doi:10.1007/s10584-007-9289.

Doesken, N. J., and A. Judson. 1996. *The Snow Booklet: A Guide to the Science, Climatology, and Measurement of Snow in the United States.* Fort Collins: Colorado State University.

Geneva Lake Environmental Agency. 2016. Newsletters. https://www.genevalakemanagement.com/geneva-waters-newsletter.

Hodgkins, G. A. 2010. Historical Ice-Out Dates for 29 Lakes in New England, 1807–2008. U.S. Geological Survey Open-Fire report 2010–2014.

IPCC. 2013. Climate Change 2013: The Physical Science Basis. Working Group I Contribution to the IPCC Fifth Assessment Report. Cambridge, UK: Cambridge University Press.

Josberger, E. G., W. R. Bidlake, R. S. March, and B. W. Kennedy. 2007. Glacier Mass-Balance Fluctuations in the Pacific Northwest and Alaska, USA. *Ann. Glaciol.* 46: 291–96.

Kocin, P. J., and L. W. Uccellini. 2005. Northeast Snowstorms. Ameri. Meteor. Soc. *Meteor. Monogr.* No. 54, 818 pp.

Kocin, P. J., P. N. Schumacher, R. F. Morales Jr., and L. W. Uccellini. 1995. Overview of the 12–14 March 1993 Superstorm. *Bull. Amer. Metero. Soc.* 76: 165–82.

Lake George Association. https://www.lakegeorgeassociation.org.

Lake Placid News. 2016. https://www.lakeplacidnews.com.

Melillo, J. M., T. C. Richmond, and G. W. Yohe, eds., 2014. Climate Change Impacts in the United States: The Third National Climate Assessment. U.S. Global Change Research Program. http://nca2014.globalchange.gov.

Minnesota Department of Natural Resources. 2019. https://www.dnr.state.mn.us/ice_out/index.html.

NASA. 2017. Antarctic Sea Ice Reaches New Record Maximum. Accessed June 2, 2019 @ https://www.nasa.gov/content/goddard/antarctic-sea-ice-reaches-new-record-maximum.

National Snow and Ice Data Center. 2014. Glacial Lake and River Ice Phenology. Accessed June 2019 @ http://nsidc.org/data/lake-river.ice.

Nenana Ice Classic. 2016. http://www.nenanaakiceclassic.com.

NOAA. 2018. Arctic Report Card Tracks Region's Environmental Changes. Accessed May 31, 2019 @ www.noaa.gov/media-release/arctic-report-card-tracks-region-s-environmental-changes.

NOAA. 2019a. Climate Change and Extreme Snow in the U.S. Accessed May 29, 2019 @ https://www.ncdc.noaa.gov/news/climate-change-and-extreme-snow-us.

NOAA. 2019b. Regional Snowfall Index (RSI). Accessed June 2, 2019 @ https://www.ncdc.noaa.gov/snow-and-ice/rsi/.

North Temperate Lakes Long-Term Ecological Research. 2016. https://lter.limnology.wisc.edu.

Ostrem, G., and M. Brugman. 1991. Glacier Mass-Balance Measurements. A Manual for Field and Office Work. Saskatoon, Sask., Environment Canada. Nation Hydrology Research Institute (NHRI Science Report 4).

Sechler, Bob, and Lori Hawkins. 2021. Winter Storm Damage May Rival Hurricane Harvey's Price Tag, Experts Say. *Austin American-Statesman*, February 19.

Smith, A. B., and R. Katz. 2013. U.S. Billion-Dollar Weather and Climate Disasters: Data Sources, Trends, Accuracy and Biases. *Nat. Hazards* 67: 387–410.

Squires, M. F., et al. 2014. The Regional Snowfall Index. American Meteorological Society. *BAMS*, 2014.

State University of New York, Oneonta Biological Field Station. 2015. Annual Reports. https://suny.oneonta.edu/biological-field-station/publications.

USEPA. 2016. Climate Change Indicators in the United Sates, 2016. Fourth Edition. EPA-430-R-16-004. www.epa.gov/climate-indicators.

USGS (United States Geological Survey). 2019a. Wolverine Glacier. https://www2.usgs.gov/landresources/lcs/glacierstudies/wolverine.asp.

USGS. 2019b. South Cascade Glacier. Accessed at https://www2.usgs.gov/landresources/lcs/glacierstudies/scascade.asp.

USGS.2019c. Gulkana Glacier. Accessed at https://www2.usgs.gov/landresources/lcs/glacierstudies/gulkana.asp.

Walters, R. A., and M. F. Meier. 1989. Variability of Glacier Mass Balances in Western North America. Accessed @ https://doi.org/10.1029/GMO55p0365.

Yukon River Breakup. 2016. Accessed June 2019. @ www.yukonriverbreakup.com.

Chapter 8

Climate Change Indicator Five

Health and Society

> Today, 1 percent of the world is too hot for humans to live. By 2070, that may increase to 19 percent.
>
> —*New York Times*, March 2021

This chapter looks at some of the ways that climate change is affecting human wellbeing—the USEPA calls this indicator heath and society. Climate change can affect public health, agriculture, water supplies, energy production and use, land use and development, and recreation. The nature and extent of these effects, and whether they will be harmful or beneficial, will vary regionally and over time.

Keep in mind that impacts of climate change on human health are complex, often indirect and dependent on multiple societal and environmental factors (including how people choose to respond to these impacts); the development of appropriate health-related climate indicators is challenging and still emerging. It is important for health-related climate change indicators to be clear, measurable, and timely to better understand the link between climate change and health effects (USEPA, 2016).

HEAT-RELATED DEATHS

Heat is the leading weather-related killer in the United States, even though most heat-related deaths are preventable through outreach and intervention. The problem stems from exposure; that is, when people are exposed to extreme heat, they can suffer from potentially deadly illnesses, such as heat exhaustion and heat stroke. High temperatures can also contribute to deaths from heart attacks, strokes, and other forms of cardiovascular disease (USEPA, 2016).

In recent decades, unusually hot summer temperatures have become more common across the contiguous 48 states, and extreme heat events (heat waves) are expected to become longer, more frequent, and more intense in the future (Melillo, Richmond, and Yohe, 2014). Accordingly, the risk of heat-related deaths and illness is also expected to increase (IPPC, 2014). Note that reductions in cold-related deaths are projected to be smaller than increase in heat-related deaths in most regions (Sarofim et al., 2016). Death rates can also change as people acclimate to higher temperatures and as communities strengthen their heat response plans and take other steps to continue to adapt.

Certain population groups already face higher risks of heat-related death, and increase in summertime temperature variability will increase that risk (Zanobetti et al., 2012; Sarofim et al., 2016). A scientific assessment has determined that the population of adults aged 65 and older, which is expected to continue to grow, has a higher-than-average risk of heat-related death. Moreover, children are particularly vulnerable to heat-related illness and death, as their bodies are less able to adapt to heat than adults, and they must rely on others to help keep them safe. People with certain diseases, such as cardiovascular and respiratory illnesses, are especially vulnerable to excessive heat exposure, as are the economically disadvantaged. Data also suggest a higher risk among non-Hispanic blacks (Berko et al., 2014).

Additional Data (USEPA, 2016)

- Between 1979 and 2014, the death rate as a direct result of exposure to heat (underlying cause of death) generally hovered around 0.5 to 1 deaths per million people, with spikes in certain years. Overall, a total of more than 9,000 Americans have died from heat-related causes since 1979, according to death certificates.

- For years in which the two records overlap (1999–2014), accounting for those additional deaths in which heat was listed as a contributing factor results in a higher death rate—nearly double for some years—compared with the estimate that only includes deaths where heat was listed as the underlying cause.
- The indicator shows a peak in heat-related deaths in 2006, a year that was associated with widespread heat waves and was one of the hottest years on record in the contiguous 48 states.
- The interaction of heat and cardiovascular disease caused about one-fourth of the heat-related deaths recorded in the "underlying and contributing causes" analysis since 1999.
- Examination of extreme events has revealed challenges in capturing the full extent of "heat-related" deaths. For example, studies of the 1995 heat wave event in Chicago suggest that there may have been hundreds more deaths than were actually reported as "heat-related" on death certificates.
- While dramatic increases in heat-related deaths are closely associated with the occurrence of hot temperatures and heat waves, these deaths may not be reported as "heat-related" on death certificates. This limitation as well as considerable year-to-year variability in the data makes it difficult to determine whether the United States has experienced a meaningful increase or decrease in deaths classified as "heat-related" over time.

DID YOU KNOW?

Classifying a death as "heat-related" does not mean that high temperatures were the only factor that caused or contributed to the death, as preexisting medical conditions can significantly increase an individual's susceptibility to heat. Other important factors, such as the overall vulnerability of the population, the extent to which people have adapted and acclimated to higher temperatures, and the local climate and topography, can affect trends in hear-related deaths. Heat response measures, such as early warning and surveillance systems, air-conditioning, health care, public education, cooling centers, infrastructure standards, and air quality management, can also make a big difference in reducing death rates (USEPA, 2016).

HEAT-RELATED ILLNESSES

Heat-related illnesses range from mild swelling, rashes, or cramps to potentially deadly heat exhaustion and heat stroke, and they cause many people to visit the emergency room or be admitted to the hospital. These illnesses can occur when a person is exposed to high temperatures, such that their body can't cool itself sufficiently through sweating.

Any person can suffer from heat stress, regardless of age, sex, or health status. Older adults and children, however, have a higher-than-average risk of becoming ill due to exposure to extreme heat (Sarofim et al., 2016). People working outdoors, the socially isolated and economically disadvantaged, those with chronic illnesses, and some communities of color are also especially vulnerable to heat (Sarofim et al., 2016).

Records indicate that unusually hot summer temperatures have become more common across the contiguous 48 states in recent decades, and extreme heat events (heat waves) are expected to become longer, more frequent, and more intense in the future (Melillo, Richmond, and Yohe, 2014). As a result, the risk of heat-related illness is expected to increase (IPCC, 2014).Then again, note that hospitalization rates can also change as people acclimate to higher temperatures and as communities strengthen their heat response plans and take other steps to continue to adapt.

Additional Data (USEPA, 2016)

- From 2001 to 2010, 20 states recorded a total of about 28,000 heat-related hospitalizations (Choudhary and Vaidyanathan, 2014).The resulting annual rates ranged from 1.1 cases per 100,000 people in 2004 to 2.5 cases per 100,000 people in 2006, with a 10-year average rate of 1.8 cases per 100,000 people.
- Heat-related hospitalization rates vary widely among the 23 states studied (Choudhary and Vaidyanathan, 2014). Average rates from 2001 to 2010 ranged from less than one case per 100,000 people in some states to nearly four cases per 100,000 people in others. The highest rates occurred in Kansas, Louisiana, Missouri, South Carolina, and Tennessee. Compared with other regions, relatively

high hospitalization rates in the Southeast and Midwest suggest a connection between hotter and more humid summers and increased rates of heat-related illness.

- People aged 65 and older accounted for more heat-related hospitalizations than any other age group from 2001 to 2010, and males were hospitalized for heat-related illnesses more than twice as often as females. Men tend to have a higher risk of heat-related illness than women because they are more likely to work in outdoor occupations such as construction (Choudhary and Vaidyanathan, 2014).

FACTORS INCREASING AN INDIVIDUAL'S RISK OF EXPERIENCING AN EXCESSIVE HEAT EVENT ATTRIBUTING TO ADVERSE HEALTH OUTCOME

Meteorological Characteristics

- Increased temperature
- Increased relative humidity
- Dry, hot winds

Demographic Characteristics

- Physical constraints (including underlying medical conditions)
- Mobility constraints
- Cognitive impairments
- Economic constraints
- Social isolation

Behavioral Choices

- Wearing inappropriate clothing
- Failing to stay adequately hydrated
- Consuming alcohol
- Engaging in outdoor activities
- Eating heavy and/or hot foods

Regional Characteristics

- Living in an area with a variable climate
- Living in an urban area
- Living on the upper floors of buildings

Source: USEPA, 2016.

DID YOU KNOW?

By any measure, summer in Phoenix, Arizona, is hot. Daily high temperatures above 100°F are routine, temperatures up to 110°F are common, and temperatures above 120°F are possible. Although residents over time physically adapt to some extent to Phoenix's high heat, summer conditions can still be quite variable. Despite Phoenix routinely experiencing life-threatening summertime temperatures, studies of excess heat mortality there have consistently found little evidence of any major heat-attributable excess mortality impacts (Kalkstein, 1997; Davis et al., 2003a,b).

HEATING AND COOLING DEGREE DAYS

This indicator examines changing temperatures from the perspective of heating and cooling needs for buildings. In accomplishing this, daily temperature data from thousands of weather stations across the United States are used to calculate heating and cooling degree days. A "degree day" is determined by comparing the daily average outdoor temperature with a defined baseline temperature for indoor comfort (in this case, 65°F); it is a way to measure the influence of temperature change on energy demand by measuring the difference between outdoor temperatures and a temperature people generally find comfortable for indoors.

How does this system work? For example, if the average temperature on a particular day is 78°F, then that day counts as 13 cooling degree days, as a building's interior would need to be cooled by 13°F to reach 65°F. On the other hand, if the average outdoor temperature is 34°F then that day counts as 31 heating degree days, as building's interior would need to be warmed by 31°F to reach 65°F. This does not mean that all people will actually heat or cool building to 65°F; it is just a number to allow for consistent comparisons over time and across the country. For reference, New York City experiences far more heating degree days than cooling degree days per year—a reflection of the relatively cool climate in the Northeast—while Houston, Texas, has far more cooling degree days than heating degree days—a reflection of the much warmer climate in the South (NOAA, 2015).

Additional Data (USEPA, 2016)

- Heating degree days have declined in the contiguous United States, particularly in recent years, as the climate has warmed. This change suggests that heating needs have decreased overall.
- Overall, cooling degree days have increased over the past 100 years. The increase is most noticeable over the past few decades, suggesting that air-conditioning energy demand has also been increasing recently.
- Heating degree days have generally decreased, and cooling degree days have generally increased throughout the north and west. The southeast, with the exception of Florida, has seen the opposite: more heating degree days and fewer cooling degree days.

VECTOR-BORNE DISEASES: LYME DISEASE AND WEST NILE VIRUS

Vector-borne diseases are those diseases that are transmitted by blood-feeding arthropods—mosquitoes, ticks, or fleas. And the purpose of this climate change indicator is to track the rate of reported disease/illness cases across the United States. Tracking and associated record keeping is vital because climate change is expected to affect the geographic and seasonal patterns of vector-borne diseases.

Lyme Disease

Lyme disease—the most common vector-borne disease in the United States—is a bacterial illness that can cause fever, fatigue, joint pain, and skin rash, as well as more serious joint and nervous system complications. In recent years, approximately 20,000 to 30,000 confirmed cases of Lyme disease per year have been reported to the Centers for

Disease Control and Prevention (CDC, 2015). Note that the actual number of illnesses is likely greater than what is reported to health officials (CDC, 2013). Lyme disease is transmitted through the bite of certain species of infected ticks (commonly referred to as deer ticks) that carry the bacteria that cause Lyme disease. These ticks live not only on deer, but also on rodents, birds, and other host animals. Deer do not harbor the bacteria that cause Lyme disease, but certain other hosts such as white-footed mice do, and the ticks pick up the bacteria by feeding on these infected hosts (USEPA, 2016).

So what part does climate play in the transmission, distribution, and incidence of Lyme disease? Studies have provided evidence that climate change has contributed to the expanded range of ticks (Beard et al., 2016), increasing the potential risks of Lyme disease. This expanded range includes areas of Canada where the ticks were previously unable to survive. The life cycle and prevalence of deer ticks are strongly influenced by temperature (Beard et al., 2016; Leighton et al., 2012). For example, deer ticks are mostly active when temperatures are above 45°F, and they thrive in areas with at least 85 percent humidity. Thus, warming temperatures associated with climate change are projected to increase the range of suitable tick habitat and are therefore one of multiple factors driving the observed spread of Lyme disease (Beard et al., 2016).

Because tick activity depends on temperatures being above a certain minimum, shorter winters could also extend the period when ticks are active each year, increasing the time that humans could be exposed to Lyme disease. Unlike some other vector-borne diseases, tick-borne disease patterns are generally less influenced by short-term changes in weather (weeks to months) than by longer-term climate change.

Changes in the populations of host species (particularly deer) and other factors affect the number of Lyme disease cases and directly affect tick population size. For example, the percentage of ticks that are infected depends on the prevalence and infection rates of white-footed mice and certain other hosts. Host species populations and habitats can be affected by climate change and other ecosystem disturbances. Human exposure to infected ticks is also influenced by multiple factors, including changes in the proximity of human populations to ticks and other hosts, increased awareness of Lyme disease, and modified behaviors, such as spending less time outdoors, taking precautions against being bitten, and checking more carefully for ticks. People who work outdoors, like landscapers and farmers, may be especially at risk (Gamble et al., 2016).

West Nile Virus

It was 1999 when **West Nile virus** was first detected in the United States. It is now the most common type of mosquito-borne disease in the United States most years. While many infected people feel no symptoms, others can experience symptoms such as headache, body aches, joint pains, vomiting, diarrhea, and rash, as well as more severe damage to the central nervous system in some patients, causing encephalitis, meningitis, and occasionally death (CDC, 2016a). From 1999 to 2014, a total of 41,762 cases of West Nile virus disease were reported to the U.S. Centers for Disease Control and Prevention (CDC). Nearly half of the reported cases were neuroinvasive—that is, affecting the brain or causing neurologic dysfunction (CDC, 2016b). Mosquitoes acquire the virus by biting infected birds, which are the main hosts of the virus. People are then infected when they are bitten by these virus-carrying mosquitoes.

Climate change may increase the risk of human exposure to West Nile virus. Studies show that warmer temperatures associated with climate change can speed up mosquito development, biting rates, and the incubation of the disease within a mosquito (Beard et al., 2016). Climate change's effects on birds may also contribute to changes in long-range virus movement, as the timing of migration and breeding patterns are driven by climate. Mild winters and drought have been associated with West Nile virus disease outbreaks (Beard et al., 2016), while breeding patterns are driven by climate. Rainfall can also contribute by creating breeding sites for mosquitoes (Beard et al., 2016).

Climate is just one of many important factors that influence the transmission, distribution, and incidence of West Nile virus disease. Human exposure to infected mosquitoes is also influenced by multiple factors, including changes in the proximity of human populations to mosquitoes and host bird species, increased awareness of West Nile virus, and modified behaviors, such as spending less time outdoors during peak mosquito-biting times and taking precautions to avoid being bitten (USEPA, 2016).

Additional Data (USEPA, 2016)

- The incidence of Lyme disease in the United States has approximately doubled since 1991, from 3.74 reported cases per 100,000 people to 7.95 reported cases per 100,000 people in 2014.
- Among the states where Lyme disease is most common, New Hampshire, Maine, and Vermont have experienced the largest increases in reported case rates since 1991, followed by Delaware and Massachusetts. On average, these five states now report 50 to 100 more cases per 100,000 people than they did in 1991.
- The incidence of West Nile virus neuroinvasive disease in the United States has varied widely from year to year. No obvious long-term trend can be detected yet.
- The years 2002, 2003, and 2012 had the highest reported incidence rates of West Nile virus, around one case per 100,000 people.
- West Nile virus occurs throughout the contiguous 48 states. Average annual incidence is highest in parts of the Southwest, the Mississippi Delta region, the great Plains, and the Rocky Mountain region.

LENGTH OF GROWING SEASON

This indicator measures the length of the growing season in the contiguous 48 states. The length of the growing season in any given region refers to the number of days when plant growth takes place. The growing season often determines which crops can be grown in an area, as some crops require long growing seasons, while others mature rapidly. Growing season length is limited by any different factors. Depending on the region and the climate, the growing season is influenced by air temperatures, frost days, rainfall, or daylight hours.

Changes in the length of the growing season can have both positive and negative effects on the yield and prices of particular crops. Overall, warming is expected to have negative effects on yields of major crops, but crops in some individual locations may benefit (IPCC, 2014). A longer growing season could allow farmers to diversify crops or have multiple harvests from the same plot. However, it could also limit the types of crops grown, encourage invasive species or weed growth, or increase demand for irrigation. A longer growing season could also disrupt the function and structure of a region's ecosystems and could, for example, alter the range and types of animal species in the area (USEPA, 2016).

Additional Data (USEPA, 2016)

- The average length of the growing season in the contiguous 48 states has increased by *nearly two weeks* since the beginning of the twentieth century. A particularly large and steady increase has occurred over the last 30 years.
- The length of the growing season has increased more rapidly in the western United States than in the east. In the West, the length of the growing season has increased at an average rate of about 2.2 days per decade since 1895, compared with a rate of nearly one day per decade in the East.
- The length of the growing season has increased in almost every state. States in the Southwest (e.g., Arizona and California) have seen the most dramatic increase. In contrast, the growing season has actually become shorter in a few southeastern states.
- In recent years, the final spring frost has been occurring earlier than at any point since 1895, and the first fall frost has been

arriving later. Since 1980, the last spring frost has occurred an average of three days earlier than the long-term average, and the first fall frost has occurred about three days later.
- Patterns in the timing of spring and fall frost largely mirror the overall patterns in the length of growing season. States that saw an increased length of growing season had comparable changes in frost-free days for both the fall and spring.

RAGWEED POLLEN SEASON

This climate change indicator gives us a picture of changes in the length of ragweed pollen season in the United States and Canada. Here's the bottom line on ragweed pollen: allergies are a major public health concern, with hay fever (congestion, runny nose, itchy eyes) accounting for more than 13 million visits to physicians' offices and other medical facilities every year (Schappert and Rechsteiner, 2011).

The culprit?

Well, consider that ragweed pollen is one of the most potent environmental allergens, which can cause hay fever and trigger asthma attacks, especially in children and the elderly (Fann et al., 2016). An estimated 15.5 percent of all Americans are sensitive to ragweed (Salo et al., 2014).

Ragweed plants mature in midsummer and produce small flowers that generate pollen. Ragweed pollen season usually peaks in late summer and early fall, but these plants often continue to produce pollen until the first frost. A single ragweed plant often continues to produce up to a billion pollen grains in one season and these grains can be carried long distances by the wind (NIAID, 2011).

Climate change can affect pollen allergies in several ways. Warmer spring temperatures cause some plants to start producing pollen earlier, while warmer fall temperatures extend the growing season for other plants, such as ragweed. Warmer temperatures and increased carbon dioxide concentrations also enable ragweed and other plants to produce more allergenic pollen, in larger quantities. This means that many locations could experience longer allergy seasons and higher pollen counts as a result of climate change (Fann et al., 2016).

Additional Data (USEPA, 2016)

- Since 1995, ragweed pollen season has grown longer at 10 of the 11 locations studied.
- The increase in ragweed season length generally becomes more pronounced in the north. Ragweed season increased by 25 days in Winnipeg, Manitoba, Canada; 24 days in Saskatoon, Saskatchewan, Canada; 21 days in Fargo, North Dakota; and 18 days in Minneapolis, Minnesota. This trend is consistent with many other observations showing that climate is changing more rapidly at higher latitudes (IPCC, 2013).
- Northern areas have seen fall frosts later than they used to, with the delay in first frost closely matching the increase in pollen season. Meanwhile, some southern stations have experienced only a modest change in frost-free season length since 1995 (Ziska et al., 2011).

REFERENCES AND RECOMMENDED READING

Beard, C. B., et al. 2016. Chapter 5: Vector-borne Diseases. The Impacts of Climate Change on Human Health in the United States: A Scientific Assessment. U.S. Global Change Research Program. http://health2016.globalchange.gov.

Berko, J., D. D. Ingram, S. Saha, and J. D. Parker. 2014. Deaths Attributed to Heat, Cold, and Other Weather Events in the United States, 2006–2010. National Health Statistics Reports, Number 76. National Center for Health Statistics. www.cdc.gov/nchs/data/nhsr076.pdf.

CDC. 2013. CDC Provides Estimate of Americans Diagnosed with Lyme Disease Each Year. www.cdc.gov/media/releases/2013/p0819-lyme-disease.html.

CDC. 2015. Lyme Disease Data and Statistics. Accessed August 2019. www.cdc.gov.lyme/stats/index.html.

CDC. 2016a. West Nile Virus Symptoms and Treatment. Accessed August 2019. www.cdc.gov/westnile/symptoms/index.html .

CDC. 2016b. West Nile Statistics and Maps. Accessed August 2019. www.cdc.gov/westnile/symptoms/index.html.

Changnon, S. A. 2007. Catastrophic Wind Storms: An Escalating Problem. *Climatic Change* 84: 131–39. doi:10.1007/s10584-007-9289.

Chestnut, L. G., W. S. Breffle, J. B. Smith, and L. S. Kalkstein. 1998. Analysis of Differences in Hot-Weather-Related Mortality across 44 U.S. Metropolitan Areas. *Environmental Science and Policy* 1, no. 1: 59–70.

Choudhary, E., and A. Vaidyanathan. 2014. Heat Stress Illness Hospitalizations—Environmental Public Health Tracking Program, 20 States 2001–2010. Surveillance Summaries 63(SS13): 1–10. www.cdc.gov/mmwr/preview/mmwrhtml/ss6313a1.htm.

Davis, R. E., P. C. Knappenberger, P. J. Michaels, and W. M. Novicoff. 2003a. Changing Heat-Related Mortality in the United States. *Environmental Health Perspectives* 111, no. 14: 1712–18.

Davis, R. E., P. C. Knappenberger, W. M. Novicoff and P. J. Michaels. 2003b. Decadal Changes in Summer Mortality in U.S. Cities. *International Journal of Biometeorology* 47, no. 3: 166–75.

Fann, N., T. Brennan, P. Dolwick, J. L. Gamble et al. 2016. Chapter 3: Air Quality Impacts. The Impacts of Climate Change on Human Health in the United States: A Scientific Assessment. U.S. Global Change Research Program.

Fountain, Henry. 2021. Global Warming's Deadly Combination: Heat and Humidity. *New York Times*, March 8.

Gamble, J. L., et al., 2016. Chapter 9: Populations of Concern. The Impacts of Climate Change on Human Health in the United States: A Scientific Assessment. U.S Global Change Research Program. https://health2016.globalchange.gov.

Hahn, M. B., et al. 2015. Meteorological Conditions Associated with Increased Incidence of West Nile Virus Disease in the United States, 2004–2012. *Am. J. Trop. Med. Hyg.* 92, no. 5: 1013–22.

IPCC. 2013. Climate Change 2013: The Physical Science Basis. Working Group I Contribution to the IPCC Fifth Assessment Report. Cambridge, UK: Cambridge University Press.

IPCC. 2014. Climate Change 2014: Impacts, Adaptation, and Vulnerability. Working Group II Contribution to the IPCC Fifth Assessment Report. Cambridge, UK: Cambridge University Press. www.ipcc.ch/report/ar5/wg2.

Kalkstein, L. S., 1997. Climate and Human Mortality: Relationship and Mitigating Measures. *Advances in Bioclimatology* 5: 161–77.

Leighton, P. A., J. K. Koffi, Y. Pelcat, L. R. Lindsay, and N. H. Ogden. 2012. Predicting the Speed of Tick Invasion: An Empirical Model of Range Expansion for the Lyme Disease Vector *Ixodes scapularis* in Canada. *J. Appl. Ecol.* 49, no. 2: 457–64.

Melillo, J. M., T. C. Richmond, and G. W. Yohe, eds., 2014. Climate Change Impacts in the United States: The Third National Climate Assessment. U.S. Global Change Research Program. http://nca2014.globalchange.gov.

Minnesota Department of Natural Resources. https://www.dnr.state.mn.us/ice_out/index.html.

National Institute of Allergy and Infectious Diseases (NIAID). 2011. Fact Sheet. www.niaid.nih.gov/topics/allergiediseases/documents/pollenallergyfactsheet.pdf.

NOAA. 2015. 1981–2010 U.S. Climate Normals. Accessed August 2019. www.ncdc.noaa.gov/data-access/land-based-station-data/land-based-datasets/climate-normals/1981-2010-normals-data.

NOAA. 2018. Arctic Report Card Tracks Region's Environmental Changes. Accessed May 31, 2019 @ www.noaa.gov/media-release/arctic-report-card-tracks-region-s-environmental-changes.

NOAA. 2019a. Climate Change and Extreme Snow in the U.S. Accessed May 29, 2019 @ https://www.ncdc.noaa.gov/news/climate-change-and-extreme-snow-us.

NOAA. 2019b. Regional Snowfall Index (RSI). Accessed Jne 2, 2019 @ https://www.ncdc.noaa.gov/snow-and-ice/rsi/.

Salo, P. M., et al., 2014. Prevalence of Allergic Sensitization in the United States: Results from the National Health and Nutrition Examination Survey (NHANES) 2005–2006. *J. Allergy Clin. Immun.* 134, no. 2: 350–59.

Sarofim, M. C., S. Saha, M. D. Hawkins, D. M. Mills et al. 2016. Chapter 2: Temperature-Related Death and Illness. The Impacts of Climate Change on Human Health in the United States: A Scientific Assessment. U.S. Global Change Research Program.

Schappert, S. M., and E. A. Rechsteiner. 2011. Ambulatory Medical Care Utilization Estimates for 2007. National Center for Health Statistics. *Vital and Health Statistics* 13, no. 169: 1–38.

USEPA. 2016. Climate Change Indicators in the United Sates, 2016. Fourth edition. EPA-430-R-16-004. www.epa.gov/climate-indicators.

Zanobetti, A., M. S. O'Neill, C. J Gronlund, and J. D. Schwartz. 2012. Summer Temperature Variability and Long-term Survival among Elderly People with Chronic Disease. *P Natl. Acad. Sci.* 109, no. 17: 6608–13.

Ziska, L., et al. 2011. Recent Warming by Latitude Associated with Increased Length of Ragweed Pollen Season in Central North America. *P Natl. Acad. Sci.* USA 108:4248–51.

Chapter 9

Climate Change Indicator Six

Ecosystems

Great Lakes Basin warming faster than other parts of country, new study finds.

—*Detroit Free Press*, headline, 2019

In this chapter we look at some of the ways that climate change affects ecosystems—ecosystems provide humans with food, clean water, and a variety of other services.

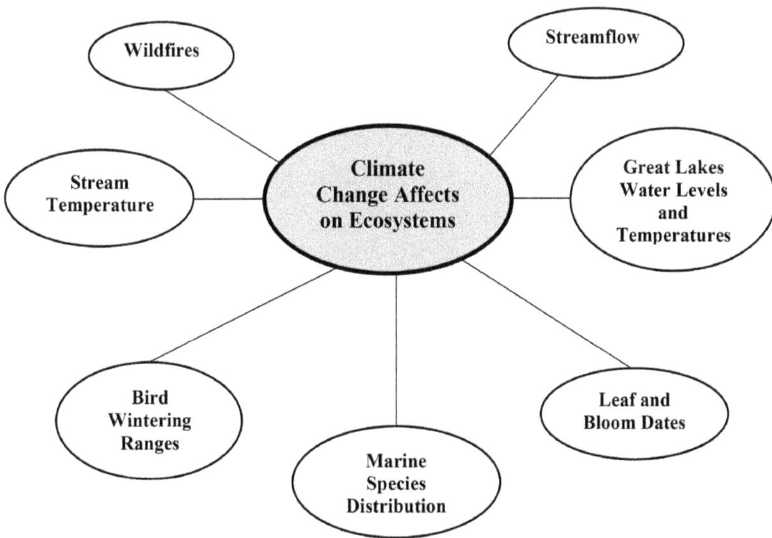

Figure 9.1. Climate Change Effects on Ecosystems

WILDFIRES

This climate change indicator tracks the frequency, extent, and severity of wildfires in the United States. Keep in mind that together, forests, shrubland, and grassland cover more than half of the land area in the United States (MRLC, 2015). These ecosystems are important resources, both environmentally and economically. While it is true that wildfires occur naturally and play an important long-term role in the health of these ecosystems, it is also true that climate change threatens to increase the frequency, extent, and severity of fires through increased temperatures and drought (USEPA, 2016). Reduced snow pack and earlier spring melting (see chapter 7) result in decreased water availability during hot summer conditions, which in turn contributes to an increased wildfire risk, allowing fires to start more easily and burn hotter. An increase in the length of the fire season has been observed in some areas (Westerling, 2016).

Other than climate change, other factors—like the spread of insects, land use, fuel availability, and management practices, including fire suppression—play an important role in wildfire frequency and intensity. All of these factors including wildfires vary greatly by region and over time, as do variations in precipitation, wind, temperature, vegetation types, and landscape conditions. Therefore, understanding changes in fire characteristics requires long-term records, a regional perspective, and consideration of many factors—like an understanding and preparing for wildfire in the wildland-urban interface (Stein et al., 2013).

Wildfires have the capability to harm property, livelihoods, and human health, particularly as population centers expand into wild land areas. The recreation and timber industries depend on healthy forests, and wildfire smoke has been directly linked to poor air quality and illness, even in communities far downwind (Johnston et al., 2012; Fann et al., 2016). Fire-related threats are increasing, especially as more people are living in and around forests, grasslands, and other natural areas (National Association of State Foresters, 2009). The United States spends around $2 billion every year to fight wildfires—in 2018 the cost was more than $3 billion (NIFC, 2020). These efforts have resulted in 480 firefighter deaths from 1990 to 2016 (NWCG, 2017).

Beyond the human impact, wildfires also affect Earth's climate. Forests in particular store large amounts of carbon. When they burn, they release carbon dioxide into the atmosphere, which in turn contributes to climate change.

Additional Data (USEPA, 2016)

- Since 1983, the National Interagency Fire Center has documented an average of 72,000 wildfires per year. Compiled data from the Forest Service suggest that the actual total may be even higher for the first few years of nationwide data collection that can be compared. The data do not show an obvious trend during this time.
- The extent of the area burned by wildfires each year appears to have increased since the 1980s. According to National Interagency Fire Center data, of the 10 years with the largest acreage burned, nine have occurred since 2000, including the peak year in 2015. This period coincides with many of the warmest years on record nationwide.
- The late 1990s were a period of transition in certain climate cycles that tend to shift every few decades (Peterson and Schwing, 2003). This shift—combined with other ongoing changes in temperature, drought, and snowmelt—may have contributed to warmer, drier conditions that have fueled wildfires in parts of the western United States (Kitzberger et al., 2007; Westerling, 2016).
- Land area burned by wildfires varies by state. Fires burn more land in the western United States than in the East, and parts of the West and Southwest show the largest increase in burned acreage between the first half of the record (1984–99) and the second half (2000–14).

STREAMFLOW

This climate change indicator describes trends in the amount of water carried by streams across the United States, as well as the timing of runoff associated with snowmelt. **Streamflow** is a measure of the rate at which water is carried by rivers and streams, and it represents a critical resource for people and the environment. Changes in streamflow can directly influence the supply of drinking water and the amount of water available for irrigating crops, generating electricity, and other needs. In addition, many plants and animals depend on streamflow for habitat and survival.

Streamflow annually varies over the course of a year. For example, rivers and streams in many parts of the country have their highest flow when snow melts in the spring and their lowest flow in late summer. The amount of streamflow is important because very high flows can cause erosion and damaging floods, while very low flows can diminish water quality, harm fish, and reduce the amount of water available for people to use. The timing of high flow is important because it affects the ability of reservoir managers to store water to meet needs later in the year. In addition, some plants and animals (for instance fish that migrate) depend on a particular pattern of streamflow as part of their life cycles.

Mechanisms that cause changes in streamflow are both natural and human-induced (USGS, 2019):

Natural Mechanisms

- Runoff from rainfall and snowmelt
- Transpiration by vegetation
- Evaporation from soil and surface-water bodies
- Groundwater discharge from aquifers
- Groundwater recharge from surface-water bodies
- Sedimentation of lakes and wetlands
- Formation or dissipation of glaciers, snowfields, and permafrost

Human-Induced Mechanisms

- Surface-water withdrawals and transbasin diversions
- River-flow regulation for hydropower and navigation

- Construction, removal, and sedimentation of reservoirs and stormwater detention ponds
- Stream channelization and levee construction
- Drainage or restoration of wetlands
- Land-use changes such as urbanization that alter rates of erosion, infiltration, overland flow, or evapotranspiration
- Wastewater outfalls
- Irrigation wastewater return flow

With regard to climate change, changes in the amount of spring snowpack and air temperatures that influence melting can alter the size and timing of high spring streamflows. More precipitation is expected to cause higher average stream flow in some places, while heavier storms could lead to larger peak flows (USEPA, 2016).

Additional Data (USEPA, 2016)

- Although most locations have experienced increases rather than decreases in streamflow during the past 75 years, there are regional differences. For example, seven-day low flows have generally increased in the Northeast and Midwest (in other words, on the days of lowest flows, streams in these areas are carrying more water than before) and low flows have generally decreased in parts of the Southeast and the Pacific Northwest (i.e., streams are carrying less water than before.
- Three-day high-flow trends vary from region to region across the country. For example, high flows have generally increased or changed little in the Northeast since 1940, whereas high flows have increased in some West Coast streams and decreased in others. Overall, more sites have experienced increases than decreases.
- Annual average streamflow has increased at many sites in the Northeast and Midwest, while other regions have seen few substantial changes. Overall, sites show more increases than decreases.
- In parts of the country with substantial snowmelt, winter-spring runoff is happening at least five days earlier than in the mid-

twentieth century at most gauges. The largest changes have occurred in the Pacific Northwest and Northeast.

STREAM TEMPERATURE

Many plants, animals, and other organisms living in streams can flourish only in a specific range of water temperatures. Temperature can affect certain aspects of water quality. For example, higher temperature reduces levels of dissolved oxygen in water, which can negatively affect the growth and productivity of aquatic life. Dissolved oxygen (DO) is the amount of oxygen that is present in water. Water bodies receive oxygen from the atmosphere and from aquatic plants. Running water, such as that of a swift moving river, dissolves more oxygen than the still water in a lake or pond.

Another problem with persistently high water temperatures in bays or streams is that natural chemical reactions can be accelerated, releasing excess nutrients into the water (Duan and Kaushal, 2013). Moreover, a stream's water temperature can also influence the circulation or mixing patterns in the water it flows into, like bays and estuaries, potentially affecting nutrient levels and salinity (USEPA, 2016).

This climate change indicator specifically discusses changes in water temperature across the Chesapeake Bay Region. We focus on the Chesapeake Bay because it is the largest estuary in the United States where it receives input from several streams, is an important habitat for countless aquatic species, and is a driver of the regional economy. Simply, water temperature of the Chesapeake Bay Region is signifi-

DID YOU KNOW?

While each aquatic organism has its own dissolved oxygen (DO) tolerance range, generally, DO levels below 3 milligrams per liter (mg/L) are of concern and waters with levels below 1 mg/L are considered **hypoxic** and usually devoid of life.

cant because it provides us with a template of the potential results of increasing water temperatures throughout the United States.

In Chesapeake Bay, as rising air temperatures cause stream temperatures to rise, warmer water coming into the bay (nine major rivers outfall or dump into the bay) can stress plants and animals. Incoming warmer water also contributes to nutrient blooms, reducing dissolved oxygen (DO) levels and creating dead zones (Duan and Kaushal, 2013).

Additional Data (USEPA, 2016)

- From 1960 through 2014, water temperature increased at 79 percent of the stream sites measured in the Chesapeake Bay region. More than half of these increases were statistically significant. Only five percent of stations had a significant temperature decrease over the same period.
- Since 1960, the Chesapeake Bay region has experienced an overall increase in stream water temperature. Temperature has risen by an average of 1.2°F across all sites and 2.2°F at the sites where trends were statistically significant.
- Stream temperatures have risen throughout the Chesapeake Bay region. The largest increases have occurred in the southern part of the region.

Trends in Stream Temperature in the Snake River (USEPA, 2017)

In addition to our focus on the Chesapeake Bay temperature increase we also include one major river, the fourth largest in the United States—the Snake River[1]—and what is going on within the River as its temperature increases. The headwaters of the Snake River are just inside Yellowstone National Park. Between 1960 and 2015, the water temperature of the Snake River increased by 1.4°F.

---

1. The Shoshone Snake people: "Yampa-pah," indicating a food plant, and "Po-og-way," meaning "road river" because it was the route followed via the Oregon Trail.

Several species of salmon use the Snake River to migrate and spawn. Salmon play a particularly important role in the diet, culture, religion, and economy of Native Americans in this region (USEPA, 2016). Climate change has challenged and will continue to challenge some of the traditional ways of life in the Snake River region. Warming river and stream temperatures threaten ecosystems and species, including salmon populations (Melillo, Richmond, and Yohe, 2014). Before we continue our discussion of temperature increases in the Snake River, it is important to include a discussion about the plight of salmon, and the many different human factors that affect salmon and those who depend on them.

Salmon are sensitive to water temperature at many stages of their lives. They spend much of their adult lives in the ocean, then—using their sense of smell as a homing device—they migrate inland to their spawning regions. Salmon need cold water to migrate and for their young to hatch and grow successfully. Warmer water can negatively affect fish making it more difficult for them to swim upstream. It can also make fish more susceptible to disease (USEPA, 2001). River and stream temperatures in the Pacific Northwest are influenced by many factors, but are expected to rise as average air temperatures increase (Cassie, 2006; Van Vliet, Ludwig, and Kabat, 2013; Isaak, Wollrab, Horan, and Chandler, 2012).

GREAT LAKES WATER LEVELS AND TEMPERATURES

This climate change indicator measures water levels and surface water temperatures in the Great Lakes. The Great Lakes (see figure 9.2), which are Lake Superior, Lake Michigan, Lake Huron, Lake Erie, and Lake Ontario, form the largest group of freshwater lakes on Earth. These lakes support a variety of ecosystems and play a vital role in the economy of the eight neighboring states and the Canadian province of Ontario, providing drinking water, shipping lanes, fisheries, recreation opportunities, and more (USEPA, 2016).

Water level and water temperature are two important and interrelated indicators of weather and climate change in the Great Lakes. Water level (the height of the lake surface above sea level) is influ-

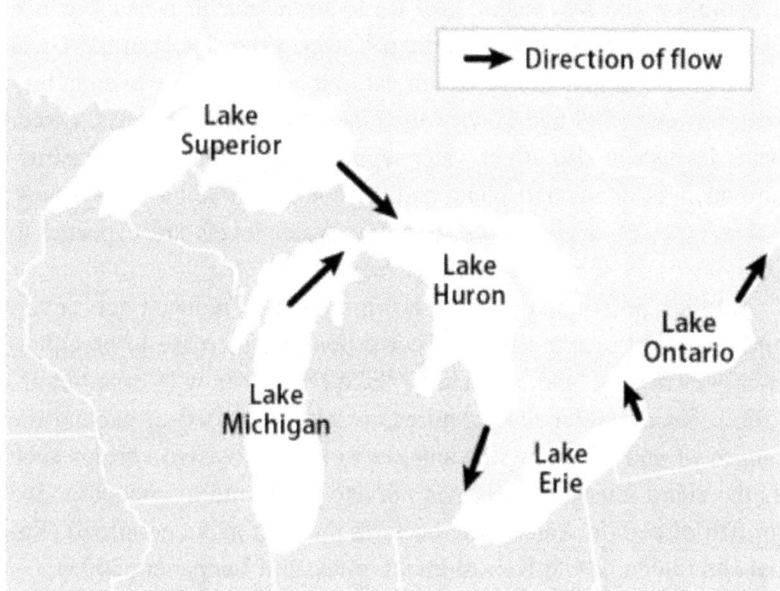

Figure 9.2. The Great Lakes
Source: USEPA (2016).

enced by many factors, including precipitation, snowmelt runoff, drought, evaporation rates, and people withdrawing water for multiple uses. Water temperature is influenced by many factors as well, but most directly by air temperature.

In recent years, warmer surface water temperatures in the Great Lakes have contributed to lower water levels by increasing rates of

DID YOU KNOW?

High pavement and rooftop surface temperatures can heat stormwater runoff. Tests have shown that pavements that are 100°F (38°C) can elevate initial rainwater temperature from roughly 70°F (21°C) to over 95°F (35°C). This heated stormwater generally becomes runoff, which drains into storm sewers and raises water temperatures as it is released into streams, rivers, ponds, and lakes. Rapid temperature changes in aquatic ecosystems resulting from warm stormwater runoff can be particularly stressful, even fatal to aquatic life (USEPA, 2016).

evaporation and by causing lake ice to form later than usual, which extends the season for evaporation. Lower water levels in the Great Lakes forced ships to reduce their cargo tonnage by five to eight percent between 1997 and 2000, which increased shipping costs. Lower water levels can also affect water supplies, the usability of infrastructure such as docks and piers, and, of course, shoreline ecosystems. These types of disturbances from low water levels are expected to continue as the climate changes (Posey, 2012).

Another potential effect of warmer water, reduced ice cover, and increased evaporation is a corresponding increase in precipitation over nearby land, especially “lake effect” snow (Burnett et al., 2003). Rising water temperatures are also expected to expand the ranges of and give new advantages to some invasive species such as the zebra mussel (*Dreissena polymorpha*), and to encourage the growth of certain waterborne bacteria that can make people ill (Rahel and Olden, 2008; Kanoshima, Urmas, and Leppanen, 2003).

THE ZEBRA MUSSEL

A hardy freshwater mussel native to Eurasia, the zebra mussel (*Dreissena polymorpha*) was probably introduced to the Great Lakes via ship ballast waters. They were first identified in the United States in the late 1980s. The problems they create are manifold: they latch onto native mussels eventually killing them, prevent native species receiving food algae, serve as a conduit to botulism that kills birds, and clog pipelines and other infrastructure that creates millions of dollars of clean-up costs for municipalities and industries that use lake water.

Additional Data (USEPA, 2016)

- Water levels in the Great Lakes have fluctuated since 1860. Over the last few decades, they appear to have declined for most of the Great Lakes. The most recent levels, however, are all within the range of historical variation.
- Since 1995, average surface water temperatures have increased slightly for each of the Great Lakes.

- Warming spring and summer months in the Great Lakes region have caused recent increases in water temperatures. These trends could relate in part to an earlier thawing of winter ice.

BIRD WINTERING RANGES

This climate change indicator examines changes in the winter ranges of North American birds. This indicator looks collectively at the "center of abundance" of hundreds of widespread North American bird species over a 48-year period. The center of abundance is a point on the map that represents the middle of each species' distribution. If a whole population of birds were to shift generally northward, one would see the center of abundance shift northward as well. Note that for year-to-year consistency, this indicator uses observations from the National Audubon Bird Count, a long-running citizen-science program in which individuals are organized by the National Audubon Society, Bird Studies Canada, local Audubon chapters, and other bird clubs to identify and count bird species. The data presented in this indicator were collected from more than 2,000 locations throughout the United States and parts of Canada. At each location, skilled observers follow a standard counting procedure to estimate the number of birds within a 15-mile diameter "count circle" over a 24-hour period. The study methods employed remain generally consistent form year to year. Data produced by the Christmas Bird Count go through several levels of review before Audubon scientists analyze the final data, which have been used to support a wide variety of peer-reviewed studies.

Birds are a particularly good indicator of environmental change for several reasons:

- Each species of bird has adapted or evolved to favor certain habitat types, food sources, and temperature ranges. In addition, the timing of certain events in their life cycles—such as migration and reproduction—is driven by cues from the environment. Consider, for example that many North American birds follow a regular seasonal migration pattern, moving north to feed and breed in the summer, then moving south to spend the winter in warmer areas. Changing conditions can influence the distribution of both migratory and nonmigratory birds as well as the timing of important life cycle events (La Sorte and Thompson, 2007).

- Birds are easy to identify and count, and thus there is a wealth of scientific knowledge about their distribution and abundance. People have kept detailed records of bird observations for more than a century.
- There are many different species of birds living in a variety of habitats, including water birds, coastal birds, and land birds. If a change in behavior or range occurs across a range of bird types, it suggests that a common external factor might be the cause (USEPA, 2016).

Analysis of recent records indicates that temperature and precipitation patterns are changing across the United States. Some bird species can adapt to generally warmer temperatures by changing where they live—for example, by migrating farther north in the summer but not as far south in the winter, or by shifting inland as winter temperature extremes grow less severe. Nonmigratory species might shift as well, expanding into newly suitable habitats while moving out of areas that become less suitable. Other types of birds might

DID YOU KNOW?

Many factors can influence bird ranges, including food availability, habitat alteration, and interactions with other species. As a result, some of the birds included in this indicator might have moved north for reasons other than changing temperatures. Keep in mind that this indicator does not show how responses to climate change vary among different types of birds. Consider, for example, that a more detailed National Audubon Society analysis found large differences among coastal birds, grassland birds, and birds adapted to feeders, which all have different abilities to adapt to temperature changes (National Audubon Society, 2009).

not adapt to changing conditions and could experience a population decline as a result. Climate change can also alter the timing of events that are based on temperature cues, such as migration and breeding (especially egg laying).

Additional Data (USEPA, 2016)

- Among 305 widespread North American bird species, the average mid-December to early January center of abundance moved northward by more than 40 miles between 1966 and 2013. Trends in the center of abundance moving northward can be closely related to increasing winter temperatures.
- On average, bird species have also moved their wintering grounds father from the coast since the 1960s. A shift always from the coasts can also relate to changes in winter temperatures. Inland areas tend to experience more extreme cold than coastal areas, but those extremes are becoming less severe as the climate warms overall (National Audubon Society, 2009).
- Note that some species have moved farther than others. A total of 48 species have moved northward by more than 200 miles. Of the 305 species studies, 186 (61 percent) have shifted their wintering grounds northward since the 1960s, while 82 (27 percent) have shifted southward. Some others have not moved at all.

MARINE SPECIES DISTRIBUTION

This climate change indicator examines changes in the location of fish, shellfish, and other marine species along U.S. coasts. It points out that changes in water temperature can affect the environments where marine species live. Moreover, this indicator tracks marine animal species in terms other than their "center of biomass," which is a point on the map that represents the center of each species' distribution by weight. If a fish population were to shift generally northward, the center of biomass would shift northward as well.

This approach is similar to the way changes in bird population distributions have been measured. One thing is clear, certain fish species naturally migrate in response to seasonal temperature changes, moving northward or to deeper, cooler waters in the summer and migrating back during the winter. As climate change causes the oceans to become warmer year-round, however, populations of some species may adapt by shifting away from areas that have become too warm and toward areas that were previously cooler. Along U.S. coasts, this means a shift northward or to deeper waters that may have a more suitable temperature. As smaller prey species shift their habitats, larger predator species may follow them (USEPA, 2016).

Marine species represent a particularly good indicator of warming oceans because of the historical record: they have been studied and tracked for several decades. Fish are especially mobile, and they may shift their location more easily than species on land because they face fewer physical barriers (Pinsky et al., 2013). Moreover, many marine species, especially fish, do not have fixed nesting places or dwellings that might otherwise compel them to stay in one place. Populations of marine species have been measured consistently for several decades across various types of ocean habitats. Tracking data for many species is useful because if a change in behavior or distribution occurs across a large range of species, it is more likely a result of a more systematic or common cause.

Data for this indicator were collected by the National Oceanic and Atmospheric Administration's (NOAA's) National Marine Fisheries Service and other agencies. These agencies monitor marine species populations by conducting annual surveys in which they trawl the ocean at regular intervals along the coast. By recording when they catch at each location, scientists can calculate each species' center of biomass in terms of altitude, longitude, and depth.

It is important to point out that water temperature is not the only factor that can cause marine animal populations to shift. Other factors could include interactions with other species, harvesting, ocean circulation patterns, habitat change, and species' ability to disperse and adapt. As a result, species might have moved northward for reasons other than, or in addition to, changing sea temperatures. This indicator is limited to the northeast and the eastern Bering Sea because these areas have used consistent survey methods and because they do not have coastlines that would prevent species from moving northward in response to warming (USEPA, 2016). *Note*: Keep in mind that data from these surveys have been used to support a wide variety of studies (Pinsky et al., 2013).

Additional Data (USEPA, 2016)

- The average center of biomass for 105 marine fish and invertebrate species shifted northward by about 10 miles between 1982 and 2015. These species also moved an average of 20 feet deeper.
- In waters off the northeastern United States, several economically important species have shifted northward since the late 1960s: American lobster, red hake, and black sea bass moved northward by an average of 119 miles.
- In the Bering Sea, Alaska pollock, snow crab, and Pacific halibut have generally shifted away from the coast since the early 1980s. They have also moved northward by an average of 14 miles.

LEAF AND BLOOM DATES

This climate change indicator examines the timing of leaf growth and flower blooms for two widely distributed plants in the United States. This climate change indicator is actually all about phenology—the study of important seasonal events and life cycles—such as the flowering of plants or the migration of birds. For example, you are observing phenology when you note when your favorite plant begins to flower each year or when your favorite migratory bird arrives at your feeder in the spring. Note that phenological events are influenced by a combination of environmental factors, including light, temperature, rainfall, and humidity. Different plant and animal species respond to different cues.

Many scientists have high confidence that the earlier arrival of spring events is linked to recent warming trends in global climate (IPCC, 2014). Disruptions in the timing of these events can have a variety of impacts on ecosystems and human society. For example, an earlier spring might lead to longer growing seasons, more abundant invasive species and pests, and earlier and longer allergy seasons. Unusually warm weather in late winter can create a "false spring" that triggers the new growth of plants too early, leaving them vulnerable to any subsequent frosts.

The timing of phenological events, due to their close connection with climate, can be used as an indicator of the sensitivity of ecological processes to climate change. Two particularly used indicators of the timing of spring events are the first leaf dates and the first bloom dates of lilacs and honeysuckles, which have an easily monitored flowering season, a relatively high survival rate, and a large geographic distribution. The first leaf date in these plants relates to the timing of events that occur in early spring, while the first bloom date is consistent with the timing of later spring events, such as the start of growth in forest vegetation (Schwartz, Ahas, and Aasa, 2006).

Additional Data (USEPA, 2016)

- First leaf and bloom dates in lilacs and honeysuckles in the contiguous 48 states show a great deal of year-to-year variability, which makes it difficult to determine whether a statistically meaningful change has taken place. Earlier dates appear more prevalent in the last few decades, however.
- Leaf and bloom events are generally happening earlier throughout the north and west but later in much of the South. This observation is generally consistent with regional differences in temperature change.

- Other studies have looked at trends in leaf and bloom dates across all of North America and the entire Northern Hemisphere. These studies have also found a trend toward earlier spring events—some more pronounced than the trends seen in just the contiguous 48 states (Schwartz, Ahas, and Aasa, 2006).

REFERENCES AND RECOMMENDED READING

Burnett, A. W., M. E. Kirby, H. T. Mullins, and W. P. Patterson. 2003. Increasing Great Lake-Effect Snowfall during the Twentieth Century: A Regional Response to Global Warming? *J. Climate* 16: 3535–42.

Cassie, D. 2006. The Thermal Regime of Rivers: A Review. *Freshwater Biology* 51: 1389–1406.

Duan, S. W., and S. S. Kaushal. 2013. Warming Increase Carbon and Nutrient Fluxes from Sediments in Streams across Land Use. *Biogeosciences* 10: 1193–1207.

Fann, N., T. Brennan, P. Dolwick, J. L. Gamble et al. 2016. Chapter 3: Air Quality Impacts. The Impacts of Climate Change on Human Health in the United States: A Scientific Assessment. U.S. Global Change Research Program.

IPCC. 2014. Climate Change 2014: Impacts, Adaptation, and Vulnerability. Working Group II Contribution to the IPCC Fifth Assessment Report. Cambridge, UK: Cambridge University Press. www.ipcc.ch/report/ar5/wg2.

Isaak, D. J., S. Wollrab, D. Horan, and G. Chandler. 2012. Climate Change Effects on Stream and River Temperatures across the Northwest U.S. from 1980–2009 and Implications for Salmonid Fishes. *Climatic Change* 113: 499–524.

Johnston, F. H., et al. 2012. Estimated Global Mortality Attributable to Smoke from Landscape Fires. *Environ. Health Persp.* 120, no. 5: 695–701.

Kanoshima, I., L. Urmas, and J. M. Leppanen. 2003. The Influence of Weather Conditions (Temperature and Wind) on Cyanobacterial Bloom Development in the Gulf of Finland (Baltic Sea). *Harmful Algae* 2: 29–41.

Kitzberger, T., P. M Brown, E. K. Heyerdahl, T. W. Swetnam, and T. T. Belen. 2007. Contingent Pacific-Atlantic Ocean Influence on Multicentury Wildfire Synchrony over Western North America. *P. Natal. Acad. Sci.* 104, no. 2: 543–48.

La Sorte, F. A., and F. R. Thompson III. 2007. Poleward Shifts in Winter Ranges of North American Birds. *Ecology* 88, no. 7: 1803–12.

Matheny, Keith, 2019. Great Lakes Basin Warming Faster than Other Parts of Country, New Study Finds. *Detroit Free Press*, March 21.

Melillo, J. M., T. C. Richmond, and G. W. Yohe, eds., 2014. Climate Change Impacts in the United States: The Third National Climate Assessment. U.S. Global Change Research Program. http://nca2014.global change.gov.

Minnesota Department of Natural Resources. Accessed June 2019. https://www.dnr.state.mn.us/ice_out/index.html.

MRLC (Multi-Resolution Land Characteristics) Consortium. 2015. National Land Cover Database 2011 (NLCD) product statistics.

National Association of State Foresters. 2009. Quadrennial Fire Review.

National Audubon Society. 2009. Northward Shifts in the Abundance of North American Birds in Early Winter: A Response to Warmer Winter Temperatures? Accessed September 2019 @ http://web4.audubon.org/bird/bacc/techreport.html.

NIFC (National Interagency Fire Center). 2020. Historical Wildland Fire Information: Federal Firefighting Costs: Suppression Only (198–2020). www.nifc.gov/fire-information/statistics/suppression-costs.

NOAA. 2015. 1981–2010 U.S. Climate Normals. Accessed August 2019. www.ncdc.noaa.gov/data-access/land-based-station-data/land-based-datasets/climate-normals/1981-2010-normals-data.

NWCG (National Wildfire Coordinating Group). 2017. NWCG Report on Wildland Firefighter Fatalities in the United States: 2007–2016. www.nwcg.gov/sites/default/files/publications/pms841.pdf.

Peterson, W. T., and F. B. Schwing. 2003. A New Climate Regime in Northeast Pacific Ecosystems. *Geophys. Res. Lett.* 30, no. 17.

Pinsky, M. L., B. Worm, M. J. Fogarty, J. L. Sarmiento, and S. A. Levin. 2013. Marine Taxa Track Local Climate Velocities. *Science* 341: 1239–42.

Posey, J. 2012. Climate Change Impacts on Transportation in the Midwest. U.S. National Climate Assessment, Midwest Technical Input Report.

Rahel, F. J., and J. D. Olden. 2008. Assessing the Effects of Climate Change on Aquatic Invasive Species. *Conserv. Biol* 22, no. 3 :521–33.

Schwartz, M. D., R. Ahas, and A. Aasa. 2006. Onset of Spring Starting Earlier across the Northern Hemisphere. *Glob. Change. Biol.* 12: 343–51.

Stein, S. M., et al. 2013. Wildlife, Wildlands, and People: Understanding and Preparing for Wildfire in the Wildland-Urban Interface. Gen. Tech. Rep. RMRS-GTR-299. Fort Collins, CO: U.S. Department of Agriculture, Forest Service, Rocky Mountain Research Station.

USEPA. 2001. Issue Paper 5: Summary of Technical Literature Examining the Physiological Effects of Temperature on Salmonids. EPA-910-D-005.

USEPA. 2016. Climate Change Indicators in the United Sates, 2016. Fourth Edition. EPA-430-R-16-004. www.epa.gov/climate-indicators.

USEPA. 2017. Tribal Connection: Trends in Stream Temperature in the Snake River. www.epa.gov/climate-indicators/snake-river.

USGS. 2016. Analysis of Data from the National Water Information System. Accessed September 2019 @ https://waterdata.usgs.gov/nwis.

USGS. 2019. Streamflow and the Water Cycle. Accessed August 30, 2019 @ https://www.usgs.gov/special-topic/water-science-school/science/streamflow-and-water-cycle.

Van Vliet, M. T. H., F. Ludwig, and P. Kabat. 2013. Global Streamflow and Thermal Habitats of Freshwater Fishes under Climate Change. *Climatic Change* 121: 739–54.

Westerling, A. L. 2016. Increasing Western U.S. Forest Wildfire Activity: Sensitivity to Changes in the Timing of Spring. *Phil. Trans R. Soc. B.* 371: 20150178.

Chapter 10

Climate Change

Challenges and Mitigations

> The recent wave of net zero targets has put the Paris Agreement's 1.5°C within striking distance.
>
> —Climate Action Tracker, December 2020

In this book's introduction, I asked, "Should we take some type of decisive action—should we come up with quick answers and put together a plan to fix these problems? What really needs to be done?"

Hopefully by this point, you have a better understanding of the elements that compose our environment (the atmosphere, the earth, the oceans—and us), the "natural" changes that have happened to Earth over the millennia, the human-made factors that complicate the environment, and the climate change indicators that govern and will govern our responses.

So, what actions can and should be taken? At what level?

There are multinational efforts—and national, state, and local efforts that can address climate change. Private industry has also made pledges. For example, for the most part the U.S. beef industry has moved from open grazing to feedlots, which gets 95 million cattle to market faster (by building up their weight faster) and thus helps reduce methane production caused by enteric fermentation. The beef industry has also changed feed components and is looking at better manure management (Fountain, 2020).

Here are some recent, news-making, climate-healing commitments:

- In September 2016, California approved Senate Bill 32 (an expansion of the California Global Warming Solutions Act of 2006), which "would require the state board to ensure that statewide greenhouse gas emissions are reduced to 40 percent below the 1990 level by 2030" (SB-32, 2016).
- In December 2020, the European Union pledged to cut net greenhouse emissions by 55 percent by 2030 (Associated Press, 2020).
- The massive multinational auto manufacturer General Motors announced in January 2021 to complete a switch to zero-emission automobiles by 2035 (and thus phase out gas-powered cars and trucks).

The key question to answer here is "What really needs to be done?" We can study the facts, the issues, the possible consequences—but the key to successfully combating these issues is to stop and seriously evaluate the problems. We need to let scientific fact, common sense, and cool-headedness prevail. Shooting from the hip is not called for or desired—makes little sense—and could have Titanic consequences for us all.

Science takes whatever this or that is and attempts to prove it—absolutely, not partially, but for sure. For example, employing the proven principles of science allows us to absolutely predict tidal flow, map out the phases of the moon, and track the effects of the changing seasons . . . and so much more. The important thing about science is that it is based on testing, factual presentations, testing, testing, and more testing. Science allows us to not only ask questions but also to answer questions about our planet and the essential aspects (physics, chemistry, and biology) of Earth.

Following are discussions of the science of two current big-scale approaches to climate change: the Paris Agreement (2015) and the proposed Green New Deal (2019) in the United States. Both of these proposals address social inequality and economic realities—these are not in the scope of this text.

THE PARIS AGREEMENT

The Paris Agreement (also known as the Paris Climate Accord) is an international treaty negotiated within the United Nations Framework Convention on Climate Change that entered into force November 4, 2016. There are currently 191 parties to the agreement (the United States left in 2020 but returned in 2021). To quote from Article 2:

> [this agreement] aims to strengthen the global response to the threat of climate change . . . by:
>
> (a) Holding the increase in the global average temperature to well below 2°C above preindustrial levels and pursuing efforts to limit the temperature increase to 1.5°C above preindustrial levels, recognizing that this would significantly reduce the risks and impacts of climate change;
> (b) Increasing the ability to adapt to the adverse impacts of climate change and foster climate resilience and low greenhouse gas emissions development, in a manner that does not threaten food production; and
> (c) Making finance flows consistent with a pathway toward low greenhouse gas emissions and climate-resilient development.

Note: as mentioned above, there are economic elements to the agreement (see section c) that are out of the scope of this book—basically, those cover financial assistance to some nations to attain the agreement's goals.

The goal of containing the global average temperature—is it achievable? This is definitely more than the $64,000 question; it is about the almost 8 billion humans that presently inhabit Earth. In the simplest terms possible we can say that the goal of containing the global average temperature is directly related to population growth. The more humans that inhabit Earth the more pollution that is created, intentional or unintentional. Also, when one country such as the United States takes climate change seriously and institutes environmental controls on the sources of pollution that is a good thing. However, the United States is low on the totem pole, relatively speaking, in regard to polluting Earth and subsequently causing global average temperature increase. In the United States, many scientific and technological advances have been made and put into practice in controlling pollution. In order to achieve the goal of containing the global average temperature it is necessary to literally take the wholistic approach; that is, it is a *global* problem and all the other inhabitants on Earth are going to have to join the United States and other nations in combatting climate change, including containing the global average temperature.

Moreover, it is important to keep in mind that population growth increases pollution on either an intentional or unintentional plane. Containing global average temperature is doable, but not easy. Only a global holistic approach based on accurate and relevant science will work.

THE GREEN NEW DEAL

The Green New Deal has come to be a title attached to many proposals around the world, and it is making an allusion to U.S. president Franklin D. Roosevelt's New Deal legislation that sought to lift the United States out of the Great Depression in the 1930s. The Green New Deal that is most newsworthy in the United States is a resolution authored by Representative Alexandria Ocasio-Cortez in 2019 that ties climate change mitigations with poverty eradication and economic justice measures (fair housing, job creation, eminent domain reforms, universal healthcare, etc.). The social policy goals of this resolution are not really applicable to this book, so I will look at climate change mitigation goals instead.

The resolution argues that "human activity is the dominant cause of observed climate change over the past century" and that global warming will cause a multitude of problems: mass migration, dangers to existing infrastructure, widespread health risks (heat stress), and more. Referencing the Paris Agreement, the resolution urges that "global temperatures must be kept below 1.5 degrees Celsius above preindustrialized levels to avoid the most severe impacts of a changing climate." To achieve that temperature goal, the climate change mitigation suggestions (to be undertaken in a 10-year period) in the 2019 Green New Deal are:

- "eliminating pollution and greenhouse gas emissions as much as technologically feasible,"
- shifting to renewable or net-zero energy sources,
- updating the power grid,
- creating green buildings through new construction or retrofitting existing buildings,
- encouraging the use of renewable energy in the manufacturing sector,
- "working collaboratively with farmers and ranchers in the United States to remove pollution and greenhouse gas emissions from the agricultural sector as much as is technologically feasible,"
- supporting "zero-emission vehicle infrastructure and manufacturing,"
- urging the creation of "clean" public transit and more high-speed rail,
- "restoring natural ecosystems through proven low-tech solutions that increase soil carbon storage, such as land preservation and afforestation,"
- "cleaning up existing hazardous waste and abandoned sites"

Other suggestions include identifying other sources of greenhouse gas emissions and sharing technology with other nations.

There's a lot going on here and a lot to do. From a scientific viewpoint, what is achievable? And will it be effective? Good questions, for sure with life-impacting implications. It is the belief of the author (and probably others) that if we honestly put our minds to accomplishing the goals listed above we can do it.

In achieving the goals listed above we need to adopt the mantra that every problem has a solution. To reach these goals we must find and incorporate the solutions—solutions based on proven science and not on feel-good science.

REFERENCES AND RECOMMENDED READING

Associated Press. 2020. EU Leaders Agree to Reduce Emissions after All-Night Talks. December 11.

Climate Action Tracker. 2020. Paris Agreement Turning Point. December 2020 update. Accessed March 2021 @ https://climateactiontracker.org/documents/829/CAT_2020-12-01_Briefing_GlobalUpdate_Paris5Years_Dec2020.pdf.

Fountain, Henry. 2020. Belching Cows and Endless Feedlots: Fixing Cattle's Climate Issues. *New York Times*, October 21.

Paris Agreement. 2015. http://unfccc.int/resource/docs/2015/cop21/eng/l09r01.pdf.

Resolution on a Green New Deal. 2019. https://ocasio-cortez.house.gov/sites/ocasio-cortez.house.gov/files/Resolution%20on%20a%20Green%20New%20Deal.pdf.

SB-32 California Global Warming Solutions Act of 2006: Emissions Limit. 2016. September 8. https://leginfo.legislature.ca.gov/faces/billTextClient.xhtml?bill_id=201520160SB32

Glossary

Acclimatization—the physiological adaptation to climatic variations.

Acid rain (pH <5.6) (in the pollution sense)—produced by the conversion of the primary pollutants sulfur dioxide and nitrogen oxides to sulfuric acid and nitric acid, respectively.

Adaptation—the adjustment in natural or human systems to a new or changing environment. Adaptation to climate change refers to adjustment in natural or human systems in response to actual or expected climatic stimuli or their effects, which moderates harm or exploits beneficial opportunities.

Adaptive capacity—the ability of a system to adjust to climate change (including climate variability and extremes), to moderate potential damages, to take advantage of opportunities, or to cope with the consequences.

Adsorption—the electrochemical attraction of positively or negatively charged molecules onto solids with an opposite charge.

Aeration—creating contact between air and a liquid by spraying the liquid in the air, bubbling air through the liquid, or agitating the liquid to promote surface absorption.

Aerobic—achieving solids reduction in manure mixtures using microorganisms that require oxygen. Thus, the breakdown of organic material tends to be odor free.

Aerosols—a collection of airborne solid or liquid particles, with a typical size between 0.01 and 10 mm that reside in the atmosphere for at least several hours. Aerosols may be of either natural or anthropogenic origin. Aerosols may influence climate in two ways: directly through scattering and absorbing radiation, and indirectly through acting as condensation nuclei for cloud formation or modifying the optical properties and lifetime of clouds.

Afforestation—the planting of new forests on lands that historically have not contained forests.

Albedo—the ratio of light reflected (reflectivity) from a particle, planet, or satellite to that falling on it.

Algal blooms—a reproductive explosion of algae in a lake, river, or ocean.

Alpine—the biogeographic zone made up of slopes above timberline and characterized by the presence of rosette-forming herbaceous plants and low shrubby slow-growing woody plants.

Anion—negatively charged ion that can adsorb to negatively charged particles. Common soil anions are nitrates and orthophosphates.

Anthropogenic—caused by humans.

Anthropogenic emissions—the emissions of greenhouse gases, greenhouse gas precursors, and aerosols associated with human activities. These include burning of fossil fuels for energy, deforestation, and land-use changes that result in net increase in emissions.

Aquifer—a stratum of permeable rock that bears water. An unconfined aquifer is recharged directly by local rainfall, rivers, and lakes, and the rate of recharge will be influenced by the permeability of the overlying rocks and soils. A confined aquifer is characterized by an overlying bed that is impermeable and the local rainfall does not influence the aquifer.

Aquitard—a geologic formation, group of formations, or part of a formation through which virtually no water moves.

Arid regions—ecosystems with less than 250 mm precipitation per year.

Atmosphere—the gaseous envelope surrounding Earth. The dry atmosphere consists almost entirely of nitrogen (78.1 percent volume mixing ratio) and oxygen (20.9 percent volume mixing ratio), together with a number of trace gases, such as argon (0.93 percent volume mixing ratio), helium, and radiatively active greenhouse gases such as carbon dioxide (0.035 percent volume mixing ratio) and ozone. In addition, the atmosphere contains water vapor, whose amount is highly variable but typically 1 percent volume mixing ratio. The atmosphere also contains clouds and aerosols.

Atmospheric stability—a property that depends on inversion strength—how rapidly air temperature rises with altitude (in units of degrees Celsius per 100 m). Strong inversions near the ground tend to stabilize the atmosphere, trap emissions, and result in higher pollutant concentrations.

Available water capacity (available moisture capacity)—the capacity of soils to hold water available for use by most plants. It is commonly defined as the difference between the amount of soil water at field capacity and the amount at wilting point. It is commonly expressed as inches of water per inch of soil.

Bar—a unit of pressure equal to one atmosphere (14.7 pounds per square inch).

Baseflow—water that having infiltrated the soil surface percolates to the ground water table and moves laterally to reappear as surface runoff.

Bedrock—the solid rock that underlies that soil and other unconsolidated material or that is exposed at the surface.

Bioaerosol—particulate matter in the atmosphere containing materials of biological origin that may cause disease, such as toxins, allergens, viruses, bacteria, and fungi.

Biodegradation—the destruction or mineralization of natural or synthetic organic materials by microorganisms.

Biodiversity—the number and variety of plants, animals, fungi, and microorganisms (such as bacteria) in a particular location.

Biomass—organic plant materials like cornstalks, small grain straw, and other plant fibers. Total amount of living material, plants and animals, above and below ground, in a particular area.

Bog—a poorly drained area rich in accumulated plant material, frequently surrounding a body of open water and having a characteristic flora (such as hedges, heaths, and sphagnum).

Boreal forest—consists of pine, spruce, fir, and larch stretching from the east coast of Canada westward to Alaska and continuing from Siberia westward across the entire extent of Russia to the European Plain.

Burden—the total mass of gaseous substance of concern in the atmosphere.

Carbon cycle—the flow of carbon (in various forms such as carbon dioxide) through the atmosphere, ocean, terrestrial biosphere, and lithosphere.

Carbon dioxide (CO_2)—a naturally occurring gas and also a by-product of burning fossil fuels and biomass, as well as land-use changes and other industrial processes. It is the principal anthropogenic greenhouse gas that affects Earth's radiative balance. It is the reference gas against which other greenhouse gases are measured.

Carbon dioxide (CO_2) fertilization—the enhancement of the growth of plants as a result of increased atmospheric carbon dioxide concentration. Depending on their mechanism of photosynthesis, certain types of plants are more sensitive to changes in atmospheric carbon dioxide concentration. In particular, plants that produce a three-carbon compound (C_3) during photosynthesis—including most trees and agricultural crops such as rice, wheat, soybeans, potatoes, and vegetables—generally show a larger response than plants that produce a four-carbon compound (C_4) during photosynthesis—mainly of tropical origin, including grasses and the agriculturally important crops maize, sugar cane, millet, and sorghum.

Carbon footprint—the amount of carbon an entity of any type (e.g., person, group, vehicle, event, building, corporation) emits into the atmosphere.

Carbon neutral—involves calculating your total climate-damaging carbon emissions, reducing them where possible, and then balancing your remaining emissions, often by purchasing a carbon offset.

Carbon offset—paying to plant new trees or investing in "green" technologies such as solar and wind power. Often these payments are a monetary donation to an environmental fund and/or endowment.

Carbon sink—anything sequestering carbon such as trees and other vegetation, forests, oceans, and grasslands.

Carbon source—anything emitting carbon into the atmosphere including forest fires, care exhaust, factories, livestock.

Carbonate—sediment formed by the organic or inorganic precipitation from aqueous solution of carbonates of calcium, magnesium, or iron.

Carbon-nitrogen ratio (C/N)—the weight ratio of carbon to nitrogen.

Catchment—an area that collects and drains rainwater.

CH_4—methane.

Chemical oxygen demand (COD)—an indirect measure of the biochemical load exerted on the oxygen content of a body of water when organic wastes are introduced into the water. If the wastes contain only readily available organic bacterial food and no toxic matter, the COD values can be correlated with BOD (biochemical oxygen demand) values obtained from the wastes.

Chlorofluorocarbons (CFCs)—greenhouse gases covered under the 1987 Montreal Protocol and used for refrigeration, air-conditioning, packaging, insulation, solvents, or aerosol propellants. Since they are not destroyed in the lower atmosphere, CFCs drift into the upper atmosphere where given suitable conditions, they break down ozone. These gases are being replaced by other compounds, including hydrochlorofluorocarbons and hydro fluorocarbons, which are greenhouse gases covered under the Kyoto Protocol.

Climate—usually defined as the "average weather" or more rigorously as the statistical description in terms of the mean and variability of relevant quantities over a period of time ranging from months to thousands or millions of years. The classical period is 30 years, as defined by the World Meteorological Organization (WMO). These relevant quantities are most often surface variables such as temperature, precipitation, and wind. Climate in a wider sense is the state, including a statistical description, of the climate system.

Climate change—a statistically significant variation in either the mean state of the climate or in its variability, persisting for an extended period (typically decades or longer). Climate change may be due to natural internal processes or external forcings, or to persistent anthropogenic changes in the composition of the atmosphere or in land use. Note that the United Nations Framework Convention on Climate Change (UNFCCC), in its Article 1, defines "climate change" as: "a change of climate which is attributed directly or indirectly to human activity that alters the composition of the global atmosphere and which is in addition to natural climate variability observed over comparable time periods." The UNFCCC thus makes a distinction between "climate change" attributable to human activities altering the atmospheric composition, and "climate variability" attributable to natural causes.

Climate forcing (see also **radiative forcing**)—"an externally imposed perturbation in the radiative energy budget of the earth climate system, for example, through changes in solar radiation, changes in the earth's albedo or changes in atmospheric gases and aerosol particles" (IPCC).

Climate model (hierarchy)—a numerical representation of the climate system based on the physical, chemical, and biological properties of its components, their interactions and feedback processes, and accounting for all or some of its known properties. The climate system can be represented by models of varying complexity—that is, for anyone component or combination of components a "hierarchy" of models can be identified, difference in such aspects as the number of spatial dimensions, the extent to which physical, chemical, or biological processes are explicitly represented, or the level at which empirical parametrizations are involved.

Climate prediction or climate forecast—the result of an attempt to produce a most likely description or estimate of the actual evolution of the climate in the future (e.g., at seasonal, interannual, or long-term timescales).

Climate system—the highly complex system consisting of five major components: the atmosphere, the hydrosphere, the cryosphere, the land surface, and the biosphere, and the interactions between them. The climate system evolves in time under the influence of its own internal dynamics and because of external forcings such as volcanic eruptions, solar variations, and human-induced forcings such as the changing composite of the atmosphere and land-use change.

Climate variability—variations in the mean state and other statistics (such as standard deviations, the occurrence of extremes, etc.) of the climate on all temporal and spatial scales beyond that of individual weather events. Variability may be due to natural internal processes within the climate system (internal variability), or to variations in natural or anthropogenic external forcing (external variability).

COD—chemical oxygen demand.

Conservative pollutants—pollutants that are not altered as they are transported from their source to the receiving water.

Contamination—the degradation of water quality as a result of natural processes and/or the activities of people.

CO_2 equivalent—the mass of carbon dioxide with the same climate change potential as the mass of the greenhouse gas in question.

Convection—the process by which air warmed by contact with the earth rises and is replaced by cold air that flows in and under it. When this cold air is warmed, it, too, rises, and is replaced by cold air.

Coral bleaching—the paling in color of corals resulting from a loss of symbiotic algae. Bleaching occurs in response to physiological shock in response to abrupt changes in temperature, salinity, and turbidity.

Cryosphere—the component of the climate system consisting of all snow, ice, and permafrost on and beneath the surface of the earth and ocean.

Deforestation—conversion of forest to non-forest.

Desert—an ecosystem with less than 100 mm precipitation per year.

Desertification—land degradation in arid, semiarid, and dry sub-humid areas resulting from various factors, including climatic variations and human activities. Further, the United Nations Convention to Combat Desertification defines land degradation as a reduction or loss in arid, semiarid, and dry subhumid areas of the biological or economic productivity and complexity of rain-fed cropland, irrigated cropland, or range, pasture, forest, and woodlands resulting from land uses or from a process or combination of processes, including processes arising from human activities and habitation patterns, such as: (i) soil erosion caused by wind and/or water; (ii) deterioration of the physical, chemical, and biological or economic properties of soil; and (iii) long-term loss of natural vegetation.

Digestion—commonly, the anaerobic breakdown of organic matter in water solution or suspension into simpler or more biologically stable compounds, or both. Organic matter may be decomposed to soluble organic acids or alcohols and then of gases such as methane and carbon dioxide. Bacterial action alone cannot complete destruction of organic solid materials.

Dispersion—the spreading and mixing of chemical constituents in ground water caused by diffusion and mixing because of microscopic variations in velocities within and between pores.

Dissolved oxygen (DO)—the molecular oxygen dissolved in water, wastewater, or other liquid; generally expressed in milligrams per liter, parts per million, or percent of saturation.

Diurnal temperature range—the difference between the maximum and minimum temperature during a day.

Drought—the phenomenon that exists when precipitation has been significantly below normal recorded levels, causing serious hydrological imbalances that adversely affect land resource production systems.

Ecosystem—a system of interacting living organisms together with their physical environment. The boundaries of what could be called an ecosystem are somewhat arbitrary, depending on the focus of interest or study. Thus, the extent of an ecosystem may range from very small spatial scales to, ultimately, the entire Earth.

Effluent—liquid discharge of a manure treatment process.

El Niño–Southern Oscillation (ENSO)—El Niño, in its original sense, is a warm water current that periodically flows along the coast of Ecuador and Peru, disrupting the local fishery. This oceanic event is associated with a fluctuation of the intertropical surface pressure pattern and circulation in the Indian and Pacific Oceans, called the Southern Oscillation. This coupled atmosphere-ocean phenomenon is collectively known as El Niño Southern Oscillation, or ENSO. During an El Niño event, the prevailing trade winds weaken and the equatorial countercurrent strengthens, causing warm surface waters in the Indonesian area to flow eastward to overlie the cold waters of the Peru current. This event has great impact on the wind, sea surface temperature, and precipitation patterns in the tropical Pacific. It has climatic effects throughout the Pacific region and in many other parts of the world. The opposite of an El Niño event is called La Niña.

Emission—in the climate change context, the release of greenhouse gases and/or their precursors and aerosols into the atmosphere over a specified area and period of time.

Energy efficiency—the ratio of energy output of a conversion process or of a system to its energy input.

Energy transformation—the change from one form of energy, such as the energy embodied in fossil fuels, to another, such as electricity.

Enteric fermentation—in ruminants such as cows, the digestive process of converting sugars into simple molecules for absorption into the bloodstream, which produces methane as a by-product.

Equivalent CO_2 (carbon dioxide)—the concentration of carbon dioxide that would cause the same amount of radiative forcing as given mixture of carbon dioxide and other greenhouse gases.

Erosion—the wearing away of the land surface by water, wind, ice, or other geologic agents and by such processes as gravitational creep.

Eutrophication—a natural or artificial process of nutrient enrichment whereby a water body becomes abundant in plant nutrients and low in oxygen content.

Evaporation—the process by which a liquid becomes a gas.

Evapotranspiration—the loss of water from an area by evaporation from the soil or snow cover and transpiration by plants.

Extreme weather event—an event that is rare within its statistical reference distribution at a particular place. Definitions of "rare" vary, but an extreme weather event would normally be as rare as or rarer than the 10th or 90th percentile. By definition, the characteristics of what is called extreme weather may vary from place to place. An extreme climate event is an average of a number of weather events over a certain period of time, an average which is itself extreme (e.g., rainfall over a season).

Forest—a vegetation type dominated by trees. Many definitions of the term forest are in use throughout the world, reflecting wide differences in bio-geophysical conditions, social structure, and economics.

Fossil CO_2 (carbon dioxide) emissions—emissions of carbon dioxide resulting from the combustion of fuels from fossil carbon deposits such as soil, natural gas, and coal.

Fossil fuels—carbon-based fuels from fossil carbon deposits, including coal, oil, and natural gas.

Front—occurs when two different air masses collide. A **cold front** marks the line of advance of a cold air mass from below as it displaces a warm air mass. A **warm front** marks the advance of a warm air mass as it rises up over a cold one.

General circulation—the large-scale motions of the atmosphere and the ocean as a consequence of differential heating on a rotating Earth, aiming to restore the energy balance of the system through transport of heat and momentum.

Glacial ablation—the loss of ice and snow from a glacier system.

Glacier—a mass of land ice flowing downhill (by internal deformation and sliding at the base) and constrained by the surrounding topography (e.g., the sides of a valley or surrounding peaks); the bedrock topography is the major influence on the dynamics and surface slope of a glacier.

Global surface temperature—the area-weighted global average of (i) the sea surface temperature over the oceans (i.e., the subsurface bulk temperature in the first few meters of the ocean), and (ii) the surface air temperature over land at 1.5 m about the ground.

Global warming—the observed warming of the planet due to human-caused emissions of greenhouse gases.

Global Warming Potential (GWP)—an index, describing the radiative characteristics of well-mixed greenhouse gases, that represents the combined effect of the differing times these gases remain in the atmosphere and their relative effectiveness in absorbing outgoing infrared radiation. This index approximates the time-integrated warming effect of a unit mass of a given greenhouse gas in today's atmosphere, relative to that of carbon dioxide.

Greenhouse effect—greenhouse gases effectively absorb infrared radiation, emitted by Earth's surface, by the atmosphere itself due to the same gases, and by clouds. Atmospheric radiation is emitted to all sides, including downward to Earth's surface. Thus greenhouse gases trap heat within the surface-troposphere system. This is called the "natural greenhouse effect." Atmospheric radiation is strongly coupled to the temperature of the level at which it is emitted. In the troposphere, the temperature generally decreases with height. Effectively, infrared radiation emitted to space originates from an altitude with a temperature of, on average, –19°C, in balance with the net incoming solar radiation, whereas Earth's surface is kept at a much higher temperature of, on average, +14°C. An increase in the concentration of greenhouse gases leads to an increase in infrared opacity of the atmosphere, and therefore to an effective radiation into space from higher altitude at a lower temperature. This causes a radiative forcing, an imbalance that can only be compensated for by an increase of the temperature of the surface-troposphere system. This is the "enhanced greenhouse effect."

Greenhouse gases—gaseous constituents of the atmosphere, both natural and anthropogenic, that absorb and emit radiation at specific wavelengths within the spectrum of infrared radiation emitted by the earth's surface, the atmosphere, and clouds. This property causes the greenhouse effect. Water vapor (H_2O), carbon dioxide (CO_2), nitrous oxide (N_2O), methane (CH_4), and ozone (O_3) are the primary greenhouse gases in Earth's atmosphere, such as the halocarbons and other chlorine- and bromine-containing substances, dealt with under the Montreal Protocol. Besides CO_2, N_2O, and CH_4, the Kyoto Protocol deals with the greenhouse gases sulfur hexafluoride (SF_6), hydrofluorocarbons (HFCs), and perfluorocarbons (PCCs).

Groundwater—water filling all the unblocked pores of underlying material below the water table.

Groundwater table—the surface between the zone of saturation and the zone of aeration; the surface of an unconfined aquifer.

Habitat—the particular environment or place where an organism or species tend to live; a more locally circumscribed portion of the total environment.

Half-life—the time required for one half of a specified substance to be transformed to another substance.

H_2S—hydrogen sulfide

Heat island—an area within an urban area characterized by ambient temperatures higher than those of the surrounding area because of the absorption of solar energy by materials like asphalt.

Hydrofluorocarbons (HFC)—one of the six greenhouse gases to be curbed under the Kyoto Protocol. They are produced commercially as a substitute for chlorofluorocarbons. HFCs largely are used in refrigeration and semiconductor manufacturing. Their global warming potentials range from 1,300 to 11,700.

Hydrosphere—the component of the climate system composed of liquid surface and subterranean water, such as oceans, seas, rivers, freshwater lakes, underground water, etc.

Hypoxia (adjective: **hypoxic**)—in ocean and freshwater environments, low or depleted oxygen in a water body (NOAA).

Ice cap—dome shaped ice mass covering a highland area that is considerably smaller in extent than an ice sheet.

Ice sheet—a mass of land ice that is sufficiently deep to cover most of the underlying bedrock topography, so that its shape is mainly determined by its internal dynamics (the flow of ice as it deforms internally and slides at its base). An ice sheet flows outward from the high central plateau with a small average surface slope. The margins slope steeply, and the ice is discharged through fast-flowing ice streams or outlet glaciers, in some cases into the sea or into ice shelves floating on the sear. There are only two large ice sheets in the modern world, on Greenland and Antarctica, the Antarctic ice sheet being divided into East and West by the Trans-antarctic Mountains. During glacier periods there were others.

Ice shelf—a floating ice sheet of considerable thickness attached to a coast (usually of great horizontal extent with a level or gently undulating surface); often a seaward extension of ice sheets.

Impacts (climate)—consequences of climate change on natural and human systems. Depending on the consideration of adaptation, one can distinguish between potential impacts and residual impacts. Potential impacts: All impacts that may occur given a projected change in climate, without considering adaptation. Residual impacts: The impacts of climate change that would occur after adaptation.

Infiltration—the process of water entering soil through the surface.

Infiltration rate—the rate at which water enters soil under a given condition, expressed as depth of water per unit time, usually inches per hour.

Introduced species—a species occurring in an area outside its historically known natural range as a result of accidental dispersal by humans (also referred to as "exotic species" or "alien species").

Invasive species—an introduced species that invades natural habitats.

Ion—a charged element or compound that has gained or lost electrons so that it is no longer neutral electrically.

kg—kilogram, 1,000 grams (about 2.2 pounds).

km—kilometer, or 1,000 meters.

Landscape—the environment, both natural and built, that surrounds us.

Leaching—the removal of soluble constituents from soils or other material by water.

Lithosphere—the upper layer of the solid Earth, both continental and oceanic, which is composed of all crustal rocks and the cold, mainly elastic, part of the uppermost mantle. Volcanic activity, although part of the lithosphere, is not considered as part of the climate system, but acts as an external forcing factor.

Mean Sea Level (MSL)—normally defined as the average relative sea level over a period, such as a month or a year, long enough to average out transients such as waves.

Methane (CH_4)—a hydrocarbon that is a greenhouse gas produced through anaerobic (without oxygen) decomposition of waste in landfills, animal digestion, decomposition of animal wastes, production and distribution of natural gas and oil, coal production, and incomplete fossil-fuel combustion. Methane is one of the six greenhouse gases to be mitigated under the Kyoto Protocol.

Microclimate—climate as experienced at the scale of particular site. Include such elements as solar orientation, wind direction, temperature, and precipitation.

Milankovitch hypothesis—in the 1920s, the Serbian astronomer Milutin Milankovitch theorized that variations/anomalies in the earth's axial tilt and orientation as it circled the sun resulted in cyclical variations in how solar radiation reached the planet. The result being changes in climate patterns over great periods of time. One explanation for ice ages.

Mitigation—an anthropogenic intervention to reduce the sources or enhance the sinks of greenhouse gases.

Mixed layer—the region of the ocean well-mixed by interaction with the overlying atmosphere.

N—nitrogen.

N_2—dinitrogen molecule.

Net carbon dioxide emissions—difference between sources and sinks of carbon dioxide in a given period and specific area or region.

NH_3—ammonia.

Nitrate nitrogen—the nitrogen component of the final decomposition product of the organic nitrogen compounds; expressed in terms of the nitrogen part of the compound.

Nitrification—oxidation of an ammonia compound to nitric acid, nitrous acid, or any nitrate or nitrite, especially by the action of nitrobacteria.

Nitrogen—a chemical element, commonly used in fertilizer as a nutrient, which is also a component of animal wastes. As one of the major nutrients required for plant growth, nitrogen can promote algal blooms that cause water body eutrophication if it runs off or leaches out of the surface soil. Nitrogen is immediately usable for plant growth in available forms.

Nitrogen cycle—the succession of biochemical reactions that nitrogen undergoes as it is converted to organic or available nitrogen from the elemental form. Organic nitrogen in waste is oxidized by bacteria into ammonia. If oxygen is present, ammonia is bacterially oxidized first into nitrite and then into nitrate. If oxygen is not present, nitrite and nitrate are bacterially reduced to nitrogen gas, completing the cycle.

Nitrogen fixation—the biological process by which elemental nitrogen is converted to organic or available nitrogen.

nm—nanometer; $^{-9}$m.

NO—nitric oxide.

NO_x—nitric oxide and nitrogen dioxide rapidly interconverted in the atmosphere.

N_2O—nitrous oxide.

North Atlantic Oscillation (NAO)—opposing variations of barometric pressure near Iceland and near the Azores. On average, a westerly current, between the Icelandic low pressure and the Azores high pressure area, carries cyclones with their associated frontal systems toward Europe. However, the pressure difference between Iceland and the Azores fluctuates on times scales of days to decades, and can be reversed at times. It is the dominant mode of winter climate variability in the North Atlantic region, ranging from central North America to Europe.

Nutrients—elements required for plant or animal growth, including the macronutrients (nitrogen, phosphorus, and potassium), which are the major nutrients required and micronutrients, which include a number of other elements that are essential but needed in lesser amounts.

Ocean conveyor belt—the theoretical route by which water circulates around the entire global ocean, driven by wind and the thermohaline circulation.

Organic aerosol—particles consisting predominantly of organic compounds, mainly C, H, and O, and lesser amounts of the other elements.

Organic matter—chemical substances of animal or vegetable origin containing carbon.

Ozone (O_3)—the triatomic form of oxygen, is a gaseous atmospheric constituent. In the troposphere it is created both naturally and by photochemical reactions involving gases resulting from human activities (photochemical "smog"). In high concentrations, tropospheric ozone can be harmful to a wide-range of living organisms. Tropospheric ozone acts as a greenhouse gas. In the stratosphere, ozone is created by the interaction between solar ultraviolet radiation and molecular oxygen (O_3). Stratospheric ozone plays a decisive role in the stratospheric radiative balance. Its concentration is highest in the ozone layer. Depletion of stratospheric ozone, due to chemical reactions that may be enhanced by climate change, results in an increased ground-level flux of ultraviolet-B radiation.

Ozone hole—see "Ozone layer."

Ozone layer—the layer in the stratosphere where the concentration of ozone is greatest. The layer extends from about 12 to 40 km. The ozone concentration reaches a maximum between about 20 and 25 km. This layer is being depleted by human emissions of chlorine and bromine compounds. Every year, during the Southern Hemisphere spring, a very strong depletion of the ozone layer takes place over the Antarctic region, also caused by human-made chlorine and bromine compounds in combination with the specific meteorological conditions of that region. This phenomenon is called the ozone hole.

PAN—peroxyacetyl nitrate.

Pathogens—disease-causing micro-organisms; generally associated with viruses or bacteria.

Percolation—the downward movement of water through soil.

Permafrost—perennially frozen ground that occurs wherever the temperature remains below 0°C for several years.

Permeability—the quality of the soil that enables water to move downward through the profile. Permeability is measured as the number of inches per hour that water moves downward through the saturated soil.

pH—a measure of the hydrogen-ion concentration. A pH of 7 is neutral; pH = 0–7 is acidic; pH = 7–14 is alkaline.

Phosphate—an ion that exists in water as H^2PO^{4-}. Otherwise phosphate is an ester or salt of phosphoric acid, such as calcium phosphate rock.

Phosphorus—one of the primary nutrients required for the growth of plants. Phosphorus is often the limiting nutrient for the growth of aquatic plants and algae.

Photochemical smog (or photochemical oxidants)—various hydrocarbons, oxides of nitrogen, and sunlight come together, initiating a complex set of reactions that produce a number of secondary pollutants.

Photosynthesis—the process by which plants take carbon dioxide (CO_2) from the air (or bicarbonate in water) to build carbohydrates, releasing oxygen (O_2) in the process. There are several pathways of photosynthesis with different responses to atmospheric CO_2 concentrations.

Phytoplankton—the plant forms of plankton (e.g., diatoms). Phytoplankton are the dominant plants in the sea, and are the best of the entire marine food web. These single-celled organisms are the principal agents for photosynthetic carbon fixation in the ocean.

Plankton—aquatic organisms that drift or swim weakly.

PM—particulate matter.

Pollutant—a resource out of place; a resource is anything useful.

Post-glacial rebound—the vertical movement of the continents and sea floor following the disappearance and shrinking of ice sheets—for example, since the Last Glacial Maximum (21 thousand years before the present). The rebound is an isostatic land movement.

Potassium—one of the primary nutrients required for the growth of plants.

ppb—parts per billion by volume.

ppm—parts per million by volume.

Precipitation—all liquid or solid phase aqueous particles that originate in the atmosphere and fall to the earth's surface.

Radiative forcing—changes in the energy balance of the earth-atmosphere system in response to a change in factors such as greenhouse gases, land-use change, or solar radiation. The climate system inherently attempts to balance incoming (e.g., light) and outgoing (e.g., heat) radiation. Positive radiative forcings increase the temperature of the lower atmosphere, which in turn increases temperatures at the Earth's surface. Negative radiative forcings cool the lower atmosphere. Radiative forcing is most commonly measured in units of watts per square meter (W/m^2).

Rapid climate change—the nonlinearity of the climate system may lead to rapid climate change, sometimes called abrupt events or even surprises. Some such abrupt events may be imaginable, such as dramatic reorganization of the thermohaline circulation, rapid deglaciation, or massive melting of permafrost loading to fact changes in the carbon cycle. Others may be truly unexpected, as a consequence of a strong, rapidly changing, forcing or a nonlinear system.

Reforestation—planting of forests on lands that have previously contained forests but that have been converted to some other use.

Renewables—energy sources that are, within a short time frame, relative to Earth's natural cycles, sustainable, and include non-carbon technologies such as solar energy, hydropower, and wind, as well as carbon-neutral technologies such as biomass.

Resource base—the combination of soil, air, water, plants, and animals that makes up the natural environment.

Runoff—the part of precipitation or irrigation water that appears in surface streams or water bodies; expressed as volume (acre-inches) or rate of flow (gallons per minute, cubic feet per second).

S—sulfur.

Salinization—the accumulation of salts in soils.

Salt—a compound made up of the positive ion of a base and the negative ion of an acid.

Saltwater intrusion/encroachment—displacement of fresh surface water and groundwater by the advance of saltwater due to its greater density, usually in coastal and estuarine areas.

Sea-level rise—an increase in the mean level of the ocean. Eustatic sea-level rise is a change in global average sea level brought about by an alteration to the volume of the world ocean. Relative sea-level rise occurs where there is a net increase in the level of the ocean relative to local land movements. Climate modelers largely concentrate on estimating eustatic sea-level change. Impact researchers focus on relative sea-level change.

Seawall—a human-made wall or embankment along a shore to prevent wave erosion.

Semiarid regions—ecosystems that have more than 250 mm precipitation per year but are not highly productive; usually classified as rangelands.

Sequestration—the process of increasing the carbon content of a carbon reservoir other than the atmosphere. Biological approaches to sequestration include direct removal of carbon dioxide form the atmosphere through land-use change, afforestation, reforestation, and practices that enhance soil carbon in agriculture. Physical approaches include separation and disposal of carbon dioxide from flue gases or from processing fossil fuels to produce hydrogen- and carbon dioxide–rich fractions and long-term storage in underground depleted oil and gas reservoirs, coal seams, and saline aquifers.

Sink—any process, activity, or mechanism that removes a greenhouse gas, an aerosol, or a precursor of a greenhouse gas or aerosol from the atmosphere.

Snowpack—a seasonal accumulation of slow-melting snow.

Solar activity—the Sun exhibits periods of high activity observed in numbers of sunspots, as well as radiative output, magnetic activity, and emission of high energy particles. These variations take place on a range of time scales from millions of years to minutes.

Solar radiation—radiation emitted by the sun. It is also referred to as shortwave radiation. Solar radiation has a distinctive range of wavelengths (spectrum) determined by the temperature of the sun.

Sorbed—adsorbed or absorbed.

Storm surge—the temporary increase, at a particular locality, in the height of the sea due to extreme meteorological conditions (low atmospheric pressure and/or strong winds). The storm surge is defined as being the excess above the level expected from the tidal variations alone at that time and place.

Stratosphere—the highly stratified region of the atmosphere above the troposphere extending from about 10 km (ranging from 9 km in high latitudes to 16 km in the tropics on average) to about 50 km.

Streamflow—water within a river channel, usually expressed in m3 sec-1.

Submergence—a rise in the water level in relation to the land, so that areas of formerly dry land become inundates; it results either from a sinking of the land or from a rise of the water level.

Subsidence—the sudden sinking or gradual downward settling of the earth's surface with little or no horizontal motion.

Sunspots—small dark areas on the Sun. The number of sunspots is higher during periods of high solar activity, and varies in particular with the solar cycle.

Sustainable development—development that meets the needs of the present without compromising the ability of future generations to meet their own needs.

Symbiotic relationship—two organisms living together in close association in which neither are harmed and both benefit.

Synthetic organic compounds—organic compounds created by industry either inadvertently as a part of a chemical process or for use in a wide array of applications for modern day life. Some that have been created are persistent in the environment (slow to decompose) because oxidizers, such as soil microbes, may not be readily able to use them as an energy source.

Tg—teragram, 1×10^{12} g.

Thermal erosion—the erosion of ice-rich permafrost by the combined thermal and mechanical action of moving water.

Thermal expansion—in connection with sea level, this refers to the increase in volume (and decrease in density) that results from warming water. A warming of the ocean leads to an expansion of the ocean volume and hence an increase in sea level.

Thermal inversion—occurs when a layer of dense, cool air is trapped beneath a layer of less dense, warm air in a valley or urban basin.

Thermohaline circulation—large-scale density-driven circulation in the ocean, caused by differences in temperature and salinity. In the North Atlantic, the thermohaline circulation consists of warm surface water flowing northward and cold deepwater flowing southward, resulting in a net poleward transport of heat. The surface water sinks in highly restricted sinking regions located in high latitudes.

Toxicity—degree of harmful effect an element or compound may have on a living organism, plant, or animal. Excessive amounts of toxic substances, such as sodium or sulfur, that severely hinder establishment of vegetation or severely restrict plant growth.

Tropopause—the boundary between the troposphere and the stratosphere.

Troposphere—the lowest part of the atmosphere from the surface to about 10 km in altitude in midlatitudes (ranging from 9 km in high latitudes to 16 km in the tropics on average) where clouds and "weather" phenomena occur. In the troposphere, temperatures generally decrease with height.

Tundra—a treeless, level, or gently undulating plain characteristic of arctic and subarctic regions.

Ultraviolet-B (UV-B) radiation—solar radiation within a wavelength range of 280-320 nm, the greater part of which is absorbed by stratospheric ozone. Enhanced UV-B radiation suppresses the immune system and can have other adverse effects on living organisms.

Urbanization—the conversion of land from a natural state or managed natural state (such as agriculture) to cities; a process driven by net rural-to-urban migration through which an increasing percentage of the population in any nation or region come to live in settlements that are defined as "urban centers."

Vector—a bearer or carrier, such as an organism (often an insect), that carries and transmits disease-causing micro-organisms.

Volatile organic compounds (VOCs)—organic molecules, usually arising from the decomposition of manure, that tend to move from liquid into the air above animal facilities.

Index

About the Author

Frank R. Spellman, PhD, is a retired assistant professor of environmental health at Old Dominion University, Norfolk, Virginia, and the author of more than 130 books covering topics ranging from homeland security to all areas of environmental science and occupational health. Many of his texts are readily available online, and several have been adopted for classroom use at major universities throughout the United States, Canada, Europe, and Russia; two have been translated into Spanish for South American markets. Dr. Spellman has been cited in more than 450 publications. He serves as a professional expert witness for three law groups and as an incident/accident investigator for the U.S. Department of Justice and a northern Virginia law firm. In addition, he consults on homeland security vulnerability assessments for critical infrastructures, including water/wastewater facilities nationwide and conducts pre–Occupational Safety and Health Administration (OSHA)/Environmental Protection Agency EPA audits throughout the country. Dr. Spellman receives frequent requests to coauthor with well-recognized experts in several scientific fields; for example, he is a contributing author of the prestigious text *The Engineering Handbook*, 2nd ed. (CRC Press). Dr. Spellman lectures on sewage treatment, water treatment, biosolids, homeland security, and safety topics throughout the country and teaches water/wastewater operator short courses at Virginia Tech (Blacksburg). He holds a BA in public administration, a BS in business management, an MBA, and an MS and PhD in environmental engineering.

Milton Keynes UK
Ingram Content Group UK Ltd.
UKHW021306151123
432621UK00023B/1230